World Fishes and Aquaculture

World Fishes and Aquaculture

George Hadwin
&
Medwin Gale

RANDOM PUBLICATIONS
NEW DELHI (INDIA)

World Fishes and Aquaculture

ISBN 978-93-5111-932-6

Published in 2016 in India by

RANDOM PUBLICATIONS

4376-A/4B, Gali Murari Lal, Ansari Road
New Delhi-110 002
Phone : +9111-43580356, 011-23289044, 011-43142548
e-mail: sales@randompublications.com,
info@randompublications.com, randomexports@gmail.com

Type Setting by : Friends Media, Delhi-110089
Printed at : Sanat Printers

Preface

World Fishes and Aquaculture" is the book of its own style which is written scholarly in easy to understand language with latest information that will fascinate student and nature buff alike.

The global commercial production for human use of fish and other aquatic organisms occurs in two ways: they are either captured wild by commercial fishing or they are cultivated and harvested using aquacultural and farming techniques.

Global production of farmed fish and shellfish has more than doubled in the past 15 years. Many people believe that such growth relieves pressure on ocean fisheries, but the opposite is true for some types of aquaculture.

The fish production from the reservoirs is low, emphasizing the need for attention to shape and develop the reservoir fisheries from the survey and planning stage to achieve high rate of production and better returns for the fishermen, who represent the weaker section of the society. Majority of these water bodies are not scientifically managed. Only a handful has so far been harnessed on scientific lines, while the others are either half-heartedly managed or even not managed at all.

Growth in aquaculture production is a mixed blessing, however, for the sustainability of ocean fisheries. For some types of aquaculture activity, including shrimp and salmon farming, potential damage to ocean and coastal resources through habitat destruction, waste disposal, exotic species and pathogen invasions, and large fish meal and fish oil requirements may further deplete wild fisheries stocks. For other aquaculture species, such as carp and molluscs, which are herbivorous or filter feeders, the net contribution to global fish supplies and food security is great3. The diversity of production systems leads to an underlying paradox: aquaculture is a possible solution, but also a contributing factor, to the collapse of fisheries stocks worldwide.

This book has been planned and written in simple and easy language, and covers the almost courses of all Indian Universities.

– *Author*

Contents

1

Status and Challenges of Fisheries

The fisheries administrators have viewed MCS as little more than the policing of the various maritime zones controlled by the State. This chapter attempts to provide a broader view of MCS - a view of MCS as the vital executive arm of fisheries management. The rapid depletion of key fish stocks in the 1980s and 1990s has made it imperative that governments achieve greater control over fishing activities. At the international level, a number of new agreements have created a stronger legal basis on which to develop greater control. At the same time, new technological developments have facilitated the remote monitoring of fishing vessels and the collection of fisheries data.

The 1982 United Nations Convention on the Law of the Sea (UNCLOS), which entered into force in 1994, forms the backbone of the international legal framework for fisheries management. It sets out the rights and duties of coastal, port and flag States in respect of each of the principal maritime zones recognized by international law, namely the territorial sea, the exclusive economic zone and the high seas. It also deals with a range of other important issues that are related, including the legal regimes applicable to internal waters, archipelagic waters, the contiguous zone, the continental shelf, and the right of innocent passage and passage through international straits. The provisions of UNCLOS relating to fisheries, though widely accepted even before the Convention entered into force, did not prevent the depletion of several valuable fish stocks. For this reason, the 1992 United Nations Conference on Environment and Development (UNCED) called urgently for the development of further instruments that would be necessary to re-establish and maintain sustainable fisheries worldwide.

One of these new instruments is the Agreement for the Implementation of the Provisions of the United Nations Convention on the Law of the Sea of 10 December 1982 Relating to the Conservation and Management of Straddling Fish Stocks and Highly Migratory Fish Stocks. The UN Fish Stocks Agreement sets forth a broad range of obligations designed to create greater control over fisheries for certain valuable stocks, including the strengthening of MCS capabilities.

The pertinent aspects of several other new instruments developed under the auspices of FAO, including:

- The 1993 Agreement to Promote Compliance with International Conservation and Management Measures by Fishing Vessels on the High Seas;
- The 1995 Code of Conduct for Responsible Fisheries (CCRF);
- Four International Plans of Action dealing with various aspects of fisheries management, and particularly the International Plan of Action to Prevent, Deter and Eliminate Illegal, Unreported and Unregulated Fishing;
- Guidelines for the marking and identification of fishing vessels.

Ideally, each State would implement the 1982 Convention and the more recent agreements through the development of a national "oceans policy." Such a policy would establish government priorities and the strategy for the conservation and sustainable use of all marine resources within the maritime zones over which the State exercises sovereignty or sovereign rights, as well as efforts towards cooperative management of fisheries that occur outside these zones. From this oceans policy would flow the integrated oceans planning and management framework under which fisheries management plans would be developed. Most States see this as a long-term development initiative. To shorten the process, they have chosen to develop oceans policy and fisheries management strategies (including MCS strategies) simultaneously. Consequently, although States recognize that fisheries management must be integrated into an overall oceans policy when it is ultimately established, the MCS systems required to implement fisheries management plans are being developed in the interim to address the more immediate need to protect fish stocks and their habitats. This strategy is commendable. A "precautionary approach" recognizes that a first requirement for fisheries resource conservation is to prevent further degradation of the resource base.

The degree to which a government becomes involved in the fishing industry will have an impact on fisheries management and the resultant MCS activities.

For example, a government can:

- Assume a controlling role, where it actually runs the industry, impacts the potential income of fishers, and micro-regulates the harvesting sector; or
- Maintain a less intrusive or co-management role, whereby the fishers and the fishing industry are encouraged to accept their resource responsibilities and roles within the framework of general government conservation principles and legislation.

Negative results of centralised, micro-management control mechanisms have become evident in both industrialised and developing States. Consequently, there is an emerging trend towards the second, *participatory co-management approach.* Fishers and the fishing industry want, and in fact are demanding, a more active role in management planning and implementation. Central governments are responding by devolving authority to smaller units of government (provinces, districts and municipalities) and by fostering community-based management, stakeholder involvement, and the acceptance of responsibility for the care, conservation and protection of their local marine resources by the fishers and industry. In a growing number of cases, such as the Canadian experience described in Profile 1, the private sector is also becoming involved in MCS activities.

MISPERCEPTIONS OF MCS

Misperceptions surrounding MCS activities continue to impact how fishers and even fishery managers view the process of MCS and enforcement. A common misperception is that all fisheries problems stem either from a failure to control illegal foreign fishing, or from the fishers themselves. While foreign fishing fleets have had documented impacts on fisheries conservation efforts, the greater impact on fisheries often stems from the domestic fishing industry in the coastal and nearshore fishing zones. On the issue of fishers, fisheries administrators must remember that most are hard working individuals often working in a hazardous environment:

While sometimes libellously assumed by the ill-informed to be crooks, [fishers] are perhaps best described as being as honest as the next man, but hard, individualistic businessmen running very competitive and often highly capitalized operations. It is worth remembering that they do so in the face of a largely unforgiving sea that creates a working environment which... [has one of the worst industrial accident rates in the world], and [that]... they operate increasingly in an economic climate of ever increasing overheads countered only by the proceeds of catches which [are] subjected to [ever greater] quota restriction. All of this is done in the knowledge that the success of their venture and the livelihood of their crews depends entirely on their individual skill, effort and initiative. Given these pressures, it is perhaps not surprising that such independent minds do not always take kindly to bureaucratic controls, especially if these appear to them to have little practical purpose. Another erroneous perception is that MCS is exclusively concerned with enforcement - thus ignoring the other two components of monitoring (data collection) and control (legislation, licensing, and controls on gear, season, areas, etc.). In focussing only on the "surveillance/enforcement" or deterrent aspects of MCS, fisheries administrators and supporting agencies cannot harness the full utility of MCS as the vital executive arm of fisheries management.

Civilian versus Military Involvement in MCS

The expense of MCS activities is often a primary concern of any government designing and implementing an MCS system. Cost-effectiveness and efficiency is important if MCS operations are to be successful. A civilian approach to deterrent fisheries enforcement has proven in many cases to be the most cost-effective and responsive to fisheries priorities. Use of civilian assets also minimizes the political sensitivity of international fisheries incidents by avoiding the use of military equipment and personnel. Those fisheries administrators who must rely on the use of military resources to carry out MCS activities may find that military agencies often accord low priority to that task. Moreover, military involvement, except in a support role, is usually not cost-effective. Military aircraft and vessels are more expensive to build and operate than equivalent civilian equipment. Savings accrue from the use of a civilian vessel with fewer crew, and lower operating costs. For many governments, however, the military can play a significant supporting role in a strong MCS system. The key for such governments is to establish an inter-agency mechanism that enables fisheries administrators to call upon their military counterparts as and when needed.

Fisheries as a Lead Ministry

Effectiveness of operations can be enhanced considerably if a single ministry is designated to take the lead role in MCS activities. This significantly reduces the lines of communications for the command and control of the monitoring and surveillance components of MCS activities, making them more efficient and responsive to management needs. As noted above, however, a number of different agencies may be called upon in a supporting role. In such situations, effective MCS requires a strong inter-agency control mechanism.

In Canada, the federal Department of Fisheries and Oceans (DFO) has been entrusted by the Parliament of Canada through the *Fisheries Act* and the *Coastal Fisheries Protection Act* to administer all laws relating to fisheries. The administration of federal fisheries laws has, by agreement, been delegated to some, but not all, provincial governments. DFO remains responsible for fisheries management in the tidal waters of the Pacific, Atlantic, Arctic, the inland waters of four Atlantic provinces and the salmon rivers in British Columbia. This includes management of Aboriginal, Recreational and Commercial fisheries within Canada's Exclusive Economic Zone (EEZ), in transboundary rivers, and for sedentary species on the continental shelf outside the Canadian EEZ.

Statistics:

- 58 400 commercial fishers (42 700 Atlantic, 8 700 Pacific and 7 000 inland)
- Commercial harvest > 1 000 000 mt
- Landed value approximately $1.9 billion (shellfish account for approximately 75 per cent of the total landed value, with groundfish and pelagic fisheries making up the remainder)

- 5 million Canadians and 900 000 visitors/year for the recreational fishery
- Catch over 250 million fish in recreational fisheries, with more than half of the fish released under the catch and release regime
- Over 125 fisheries agreements are negotiated with Aboriginal groups in Canada on an annual basis to provide for aboriginal access to fisheries and an orderly management of their fishing activities.

Management Systems/Control Mechanisms

Various management schemes have been applied in different fisheries, but all are based on limited entry licensing, with vessel and gear restrictions. Other measures to limit catches include total allowable catches (*e.g.* groundfish), escapement targets (salmon), or recruitment strategies (*e.g.* lobster). Other management measures include limitations on fishing area, fishing season, gear (*e.g.* mesh size), incidental catch (bycatch) and minimum fish sizes. Rights-based systems - in the form of Enterprise Allocations (EA) and Individual Quotas (IQ) - have been introduced in some fisheries to allow fishers more efficiently to manage the capacity and effort for harvesting.

MCS Programme

The Department of Fisheries and Oceans (DFO) Conservation and Protection programme ensures compliance with the legislation, regulations and fishing plans. This requires an integrated MCS approach and deployment of some 600 Fishery Officers for air, sea and land patrols; independent/private sector observer coverage on fishing vessels; dockside monitoring of fish landings; and remote electronic monitoring of fishing vessel activity.

DFO operates a fleet of patrol vessels on each coast to enforce closed areas and boundary lines, and to conduct inspections at sea for compliance purposes. Contracted aircraft are used to monitor fishing fleets. Sea and aerial surveillance is also supplemented by the Department of National Defence (DND).

VMS and air surveillance provide for more effective deployment of patrol vessels; with the latter also serving as a visible deterrence. Canada deploys private sector, contracted observers, without enforcement powers, on all foreign vessels fishing in Canadian waters and on some Canadian vessels to gather scientific information and provide on-site monitoring of compliance. They are trained to detect and report infractions such as dumping/discarding, fishing in closed areas, catch misreporting, retention of prohibited catch and the use of illegal gear.

The level of observer coverage in domestic fisheries varies depending on conservation risks and management priorities. Contracted dockside monitors/ observers verify the quantity and species/product form of fish landed for scientific, quota monitoring and compliance reasons. This data is cross-checked by random inspections by Fishery Officers at landing sites. At-sea observer costs are shared by the Department and the fishing industry. Dockside monitoring costs are entirely paid by the fishing industry. In 1999, DFOs expenditures on fisheries enforcement amounted to CDN $70 million (including DFOs share of at-sea observer costs).

The Conservation and Protection programme has been significantly re-oriented in recent years. The mix of enforcement resources has been altered to better respond to changing programme requirements. For example, a number of larger patrol vessels have been replaced with smaller programme boats that can be operated more efficiently by the Fishery Officers themselves. Savings from vessel reductions have been partially re-invested in new equipment and surveillance technologies. Significant investments have also been made in the creation of new enforcement data systems and the integration of existing systems, with the goal of providing Fishery Officers and managers with more accurate and timely information that will strengthen the Department's enforcement capabilities. These efforts at improving data integration and analysis will continue to be a priority for the immediate future. The Conservation and Protection programme is closely integrated with DFOs overall Fisheries Management Programme. Input and advice from fisheries enforcement officials is an important consideration in the development of Integrated Fisheries Management Plans (IFMPs). In future these plans will include specific conservation objectives as identified by DFO scientists in consultation with fishers and other technical experts. The new Objectives Based Fisheries Management strategy will require greater involvement by the fishing industry to design management measures that will minimize identified risks to conservation.

Conserve and Utilise Marine resources

Fisheries are critical to the development of a State's plan to conserve and utilise marine resources, as fish and their habitat are significant renewable resources in the territorial sea and exclusive economic zone. The goal of fisheries management, including MCS, is to maximise the economic opportunities and benefits from the State's waters within sustainable harvesting limits. Fisheries MCS needs to be defined in light of this goal. An MCS Conference of Experts organized by FAO in 1981 developed a definition of MCS that is commonly accepted by fisheries personnel:

- Monitoring - the continuous requirement for the measurement of fishing effort characteristics and resource yields;
- Control - the regulatory conditions under which the exploitation of the resource may be conducted; and
- Surveillance - the degree and types of observations required to maintain compliance with the regulatory controls imposed on fishing activities.

Simply stated, MCS is the mechanism for implementation of agreed policies, plans or strategies for oceans and fisheries management. MCS is an aspect of oceans and fisheries management that is often undervalued. In reality, it is key to the successful implementation of any planning strategy. The absence of MCS operations render a fisheries management scheme incomplete and ineffective. Since the 1981 MCS Conference, the definition of MCS has been enhanced to promote the concept that MCS covers more than just fisheries enforcement - it is an integral and key component for the implementation of fisheries management plans. It encompasses not only traditional enforcement activities but also the development and establishment of both data collection systems, the enactment of legislative instruments and the implementation of the management plan through participatory techniques and strategies.

A 1993 workshop in Ghana offered the following clarifications:

- *Monitoring* includes the collection, measurement and analysis of fishing activity including, but not limited to: catch, species composition, fishing effort, bycatch, discards, area of operations, etc. This information is primary data that fisheries managers use to arrive at management decisions. If this information is unavailable, inaccurate or incomplete, managers will be handicapped in developing and implementing management measures.
- *Control* involves the specification of the terms and conditions under which resources can be harvested. These specifications are normally contained in national fisheries legislation and other arrangements that might be nationally, subregionally, or regionally agreed. The legislation provides the basis for which fisheries management arrangements, via MCS, are implemented. For maximum effect, framework legislation should clearly state the management measures being implemented and define the requirements and prohibitions that will be enforced.
- *Surveillance* involves the regulation and supervision of fishing activity to ensure that national legislation and terms, conditions of access, and management measures are observed. *This activity is critical to ensure that resources are not over exploited, poaching is minimized and management arrangements are implemented.*

These wider definitions amplify the importance of all aspects of MCS.

RESPONSES TO THE CHALLENGES OF FISHERIES USE

The very reality that the root causes of the over-use of fish stocks are structural makes it nearly impossible to completely alleviate the pressures of rising demand, increasing abilities to pay for high-value species, and more effective ways to capture fish on an open-access resource. But people, organizations, and governments have responded to these realities in many ways, their responses sometimes improving the situation, but often exacerbating it.

Two of the most prominent and important responses have been subsidies and management schemes.

The first, subsidies, occur when trade measures directly influence the income or reduce the costs of production for industry. Management schemes, the second, encompass the ways in which fisheries are governed and monitored. In practice, the way this is done often highlights a lack of effective management. Subsidies and management, which can be seen as responses by individuals, groups, or governments to the root causes of fisheries depletion, are also the areas where the most opportunities for change in the way fisheries resources are used exist.

Effective management regimes are considered by many to be key in the sustainable use of fisheries resources. Theoretically, effective management encourages a sustainable use of resources through providing conservation regulations and incentives for responsible use of the resource. They are especially beneficial under the pressures of international integration; responses by producers to the economic incentives from trade liberalization, for example, can be better regulated when effective management schemes exist so that over-exploitation and destruction are less likely to occur. Unfortunately, the prevalence of such management has been scant in real-world fisheries.

As the United Nations Environment Programme (UNEP) reports, "very few fisheries management systems have demonstrated the ability to keep catches below levels that put pressure on stocks." As noted in a study by the World Conservation Union (IUCN), "national and international progress on trade expansion and liberalization outpaces progress on fisheries management and articulation of sustainable development strategies." Fisheries management, crucial to sustainable use of the world's fisheries, has been difficult to implement effectively and has often left fisheries open to influences from other sources such as state trade measures.

Subsidies, one of these trade measures, have played a proven role in aggravating the causes of fisheries depletion. Theoretically, subsidies can lead to overuse of a resource by altering the behaviour of producers. Revenue-enhancing or cost-reducing subsidies, such as those that will be examined in this chapter, increase marginal profits at each level of effort and therefore theoretically lead to an increased overall effort. This may have the short-term effect of creating additional economic rent for producers, but either new entrants or increased effort by existing producers, also encouraged by the subsidy, will shift the level of effort so that rent is eventually dissipated. Assuming that the management programme does not effectively impose a sustainable level of resource use, cost-reducing and revenue-enhancing subsidies will push the level of overcapacity and overall effort even further than would an open-access resource in the absence of subsidies. The main theoretical effects of cost-reducing, rent-increasing subsidies are increased capacity, a delay in exit by producers from the sector, and increased effort by producers. These effects can cause the overuse of a resource, which can have many other significant impacts such as threats to food, livelihood, and resource security.

FISHING SUBSIDIES

In the real world of governments, fishers, fisheries resources, and the incentives that link them all together, a subsidy can take many forms. Fishing subsidies currently account for large portions of many industrialized country budgets; a 1998 World Bank study estimated annual budgeted expenditures on global fisheries subsidies to be between $14 and $20 billion, and that subsidies account for 20 to 25 per cent of the sector's revenues.

One type of subsidy, domestic assistance programmes, form a clear example of government endorsement of the fishing sector. These programmes are defined as budgeted assistance to fishing production, and often enhance operations and capacity. Japan, one of the biggest subsidizers of fishing activities in the world, has used domestic assistance programmes to recruit young fisherman, give aid to fish cooperatives and boat owners, support marketing, and encourage price stabilization. A 1998 World Bank report estimated that Japan budgets $270 million a year for these types of programmes, and a total of $750 when aid for fishing vessel insurance is included. The European Union (EU), another major subsidizer, supports such structural programmes as fleet renewal and modernization, processing and marketing, aquaculture, maintenance of port facilities, and genetic product promotion in their fisheries sector.

In addition, it pays for market supports to the sector in the form of a minimum import price programme and measures to support price floors. Costs of these programmes to the EU in 1996 were estimated to be about $530 million. While estimates of exactly what types of domestic subsidy programmes the Chinese government uses are hard to determine due to scant data, they have clearly aggressively promoted the expansion of their fishing sector in the last 15 years (and have become the world's leading producer) and are estimated to budget between $500 and $750 million annually for domestic subsidies. While other countries subsidize their domestic fishing industries, they have comparatively small budgets for such programmes as so do so on a smaller scale. The shear magnitude of the domestic subsidies in industrialized countries suggests the possible impacts on the behaviour of fishers their presence may cause.

A less obvious, but yet potentially just as influential, type of subsidy occurs when a country pays the access fees for their fleets to fish in foreign waters. Unlike domestic subsidies, these payments are not made directly to fishers or even domestic programmes, but to the foreign governments who have jurisdiction over the fish stocks. The EU is one of the major subsidizers of distant-water fishing through its numerous access agreements with developing countries, mostly in West Africa.

For example, the EU signed a five-year access agreement with Mauritania in 1996 that lifted an EU embargo on fishery imports from Mauritania and specified EU payments of almost $350 million over five years to the Mauritanian government. In return, the Mauritanian government granted increased access for EU fishing vessels in their waters, authorization of higher EU total harvests, and the specific allowance of EU-directed fishing of highly valued squid and octopus.

The monetary benefits to the "cash-strapped" nation were described as a "windfall;" in consequence, the number of eligible EU boats rose from 165 to 240 and allowable EU harvests from 76,050 to 183,392 tons. The EU has similar agreements with other West African nations, as well as others in East Africa, the Indian Ocean, and South America. Another distant-water fleet subsidizer, Japan, currently budgets about $200 million a year on distant water arrangements with foreign countries and foreign fisheries assistance programmes (which often help secure fishing rights). These payments went mostly to developing countries in the Pacific Ocean and elsewhere. The subsidization of distant water fleets from industrialized nations can have serious effects on the coastal communities of the developing countries with whom these agreements are typically made.

Possible Impacts of Perverse Subsidization

The impacts of government subsidies of fishing are at least as complex and numerous as the types of subsidies themselves, but in the context of sustainable development those related to market access due to trade measures and resource, livelihood, and food security are the most relevant. The trade implications of the subsidization of the fishing industry arise when the lower costs producers see as a result of the subsidies repress world prices for fish products.

Those producers, who are often from developing countries that cannot afford subsidies, are then excluded from the market because they cannot afford to produce at the world price.

For example, when Japan subsidizes its fleets to fish in the waters of Chile, Japanese fishers bring the fish back to Japan and sell them on their domestic market. Chilean fishers try to enter the Japanese market and sell these same Chilean fish to Japanese consumers, but are kept out of the market because their costs are higher and the prices on the Japanese market are too low. In effect, Japanese fishers push Chilean producers out of the market for the sale of their own fish. This example illustrates the way in which subsidies in the fishing sector can exclude those who are outside the national subsidy regime from participating in the market.

Due to numerous examples where the presence of subsidies seemed to increase the capacity in the sector and, in turn, deplete available resources, fishing subsidies have also been blamed for a tendency to encourage overcapacity and overfishing.

The case of stock depletion in Canada's Northwest Atlantic fisheries in the 1950s and 1960s illustrates how subsidies can increase capacity and lead to depletion. Between 1954 and 1968, Canadian subsidies increased the capacity of the Northwest Atlantic offshore fishing fleet by more than 18 times, creating twice as much capacity as could capture fish if resources were to be used sustainably. In 1970, the Canadian government "acknowledged explicitly that its subsidies for vessel construction and modernization over the previous two decades had led to the rapid expansion of larger vessels, creating serious overcapitalisation." An Organization for Economic Cooperation and Development (OCED) study reported that subsidies also contributed to overcapacity in the fishing sector in many member states: New Zealand in the 1970s and early 1980s, Spain's Galician fisheries in the 1980s, Norway in the 1960s, the EU during the 1980s, and the United States (US) in the late 1970s and 1980s. These examples of the history of fishing fleet subsidization in industrialized countries and the resulting depletion of their stocks illustrates how overcapacity in a sector can encourage overuse of a fragile resource and lead to its exhaustion, posing threats to the surrounding ecosystem and human environment.

In addition to threatening the resources on which humans depend for their incomes, fishing subsidies may pose direct threats to crucial aspects of human and national development such as livelihood and food security. These threats are seen most strikingly in the interaction between developing nations and the industrialized countries with whom they sign access agreements. Frequent realities of EU-West African access agreements, for example, include a disregard of African fishing regulations and international agreements by distant water fleets, an inability by African states to monitor activities or catch levels of EU fleets, a lack of direct linkage between access payments and money spent on fisheries-related activities, a lack of enforcement of management efforts by African governments due to a fear of losing compensations, and the challenge of depletion of valuable fisheries stocks.

There are also often many adverse effects on African coastal communities such as threats to the livelihoods of artisanal fishers and food security of those who depend of local fish stocks for protein intake. While EU payments often constitute important sources of revenue for West African governments, the security of fishing resources, vital to so many aspects of these countries, are often threatened.

The case of one West African country that signed an access agreement with the EU, Senegal, illustrates how subsidies and trade agreements can lead to threats on food and livelihood security. The Senegalese government initially implemented subsidies as a way to encourage fish production to supply domestic consumers with a protein source; fish accounts for 75 per cent of the population's protein needs. But the use of subsidies has since become incompatible with this goal since the end result has been a re-orientation towards encouraging the exportation of fisheries resources and agreements with the EU for access to Senegalese fisheries.

The combination of domestic export subsidies, which lower the costs of production for those fishers who chose to export, and trade advantages granted under the Lomé Agreement of 1982, which authorized customs exemptions to products originating in African, Caribbean, and Pacific (ACP) countries and increased the competitiveness of Senegalese piscatorial products in the European market, have encouraged small-scale fishers in the country to both intensify their efforts in response to international demand as well as orient themselves towards the export market.

These developments in subsidy policy, along with several fisheries agreements that have been signed with Japan and the EU, have threatened livelihood and food security in Senegal.

The impact of growth in subsidy-driven exports has lead to a "breakdown in the supply of cheap protein to the population," since local producers began to switch their activities towards the capture of species of high market value in European economies rather than those affordable to local consumers. The increasing scarcity of fisheries resources has also affected the traditional, artisanal fishers, whose techniques, which require low technology and employ many people, are beneficial for local resource and employment security. Since fish have become scarcer, foreign fishing fleets have become a larger presence, and local fishers lack the equipment and fuel to travel farther out to sea where the remaining catches are located, they have been forced to make deals with European or Asian boats to remain employed.

In return, locals allow the foreign boats access to areas originally reserved for their needs and use. This trend threatens local control of the resource as well as livelihood security. In a country where fish provides a vital source of sustenance for the local population, in terms of both food and employment, the way in which subsidies encourage a reorientation of resources to foreign markets is a matter of much concern for the sustainable development of the country.

CAPABILITIES OF DEVELOPING COUNTRIES IN FISHERIES

Clearly, the needs and capabilities of developing countries in fisheries vary widely. In this report we have identified the main strategic issues and trends in fisheries and aquaculture from a global perspective. The strategy that we advocate for USAID is to identify and address the common needs within and among countries to benefit the greatest number of people for the resources available.

Considering the issues and future trends we have identified in this study, we recommend the following strategic approaches to USAID:

USAID needs to substantially increase its programmatic emphasis and enlarge its *financial and human resource commitments to global fisheries and aquaculture.*

Given the importance of fish and fisheries to the global economy and their importance in poverty alleviation and food security, USAID has ample justification to increase its global profile in fisheries and aquaculture. We strongly recommend that USAID substantially increase its programmatic emphasis, funding, and enlarge its staff commitments to global fisheries and aquaculture issues.

If USAID cannot increase its core commitments in these areas, the Agency should consider augmenting its core staff by rotating into USAID mid-and senior level professionals from state and federal governments, academia, NGOs, and industry.

USAID needs to make more prominent the importance it gives to the sustainability of fisheries and aquaculture at all levels of its bureaucracy: at its DC headquarters, and in all of its missions, and its regions. In addition, USAID needs to give a higher profile to capture fisheries and aquaculture issues in descriptions of its overall agriculture and natural resources portfolios. We find the word "fisheries" missing from these; and we believe aquaculture is too "buried" in lists describing a wide range of agricultural issues that interest the USAID.

USAID needs to play a central role in mobilizing America's considerable human *and institutional resources in fisheries and aquaculture to assist developing countries.*

Our vision is one in which the USAID is a world leader in channeling high quality, "needs directed" kinds of technical assistance in fisheries and aquaculture to developing countries, mainly in the form of capacity building though education and training opportunities, but also in applied research. In order to achieve this vision, we believe that the USA needs to increase its competence and competitive position in dealing with fisheries and aquaculture issues in developing countries.

USAID needs to play a central role in mobilizing America's considerable human and institutional resources in fisheries and aquaculture to assist developing countries. This can be brought about by a variety of initiatives, but the basic requirement is for US applied science and outreach professionals to work more often and for longer durations in developing country fisheries and aquaculture situations in a way that focuses more closely on collaborative solutions to common issues.

A key element of USAID approaches must be the leveraging its organizational and leadership strengths in fisheries and aquaculture by utilizing the considerable array of expertise and talent that exists in America's government (USDA, NOAA-OAR, NOAA-Sea Grant, NOAA-Fisheries, USGS, etc.), universities, state agencies, industry, and NGOs. We encourage USAID to develop innovative ideas/proposals to link with the interests of the US State Department and the Peace Corps to develop additional policy, research, and extension capabilities.

In addition, the USAID needs to better insure that its multi-lateral and bilateral investments make the most use of American expertise in fisheries and aquaculture, especially in regards to instruments and relationships with The World Fish Center, various UN organizations, The World Bank, and The Asian, African, and Interamerican Development Banks.

The Pond Dynamics/Aquaculture CRSP (PD/A CRSP) is one notable example of how the USAID can play an enhanced role in mobilizing America's considerable human and institutional resources in fisheries and aquaculture to assist developing countries. The PD/A CRSP has been involved with over 50 institutions and NGOs in 27 host countries. Since the inception of the PD/A CRSP over 200 researchers have been involved and over 400 graduate and undergraduate students supported.

During the past eight years the PD/A CRSP has worked (or is currently working) with 18 US universities in 16 states. This vital capacity-building has enhanced and strengthened host countries' abilities to further develop aquaculture and provide an additional and much needed protein source to the local and regional populations.

However, USAID funds have been far too limited to develop long-term University centers of excellence in capture fisheries and aquaculture; plus the CRSP lacks a broader mandate in order to engage fully in the urgent issues of nearshore and inland fisheries, marine aquaculture, coastal area management, and the comprehensive, systems-based natural resource management approaches we recommend that would ensure a sustainable future for capture fisheries and aquaculture.

The USAID needs to examine how the CRSP, the USAID Cooperative Agreements, and other current/past administrative initiatives in coastal management and aquaculture could be used as models to expand long-term engagement in critical regional issues that have place-based centers and can involve an expanded collaboration between USAID/Government/Universities that could build on America's strengths and competitive position.

Foreign nationals need specialized short and long-term training in the US. Trained human resources are essential to resource management, which require, inter alia: multi-disciplinary expertise in fishery resource assessment; bioeconomic and socio-economic analysis; management techniques; fishing technology, marketing and quality control; resource monitoring; fishery surveillance; and fisheries legislation.

The USAID Cooperative Agreements in aquaculture, fisheries and comprehensive coastal area management with American universities were very effective programmes for capacity building worldwide. These Agreements created an impressive cadre of globally important leaders—and good will—in fisheries and aquaculture throughout the academic and governmental institutions of many nations.

We urge USAID to enhance its commitments to the building of additional leadership capacities in developing nations by elevating overall human resource capabilities to better manage fisheries and aquaculture. Additional Cooperative Agreements for training on a variety of concepts and skills are needed in order to have a chance at sustaining and rehabilitating fisheries and aquaculture ecosystems in many nations, to strengthen institutions, and to improve individual performance.

We recommend USAID to:

- Increase funding to and the participation of American scientists in The World Fish Center, and more generally, in the current and planned activities in fisheries and aquaculture of the Consultative Group in International Agriculture Research (CIGAR),
- Participate actively in the UN/FAO Associate Professional Officer programme, so that young professionals can gain varied and broad global experiences in international settings,
- Increase the recruitment of fishery and aquaculture graduates into USAID missions and train them more extensively in pre-service at American universities,
- Fund or facilitate additional foreign student degree and certificate programmes in fisheries and aquaculture and associated resource sciences at US institutions (both at universities and at US government organizations),
- Fund or facilitate additional targeted technical assistance missions (social science, economics, fisheries management, processing, labeling, GIS, HACCP, etc.),

We recommend that USAID lead a planning process that could result in establishment of formal collaborations with a suite of US government/University centers of excellence— possibly using an expanded and better funded CRSP mechanism—in order to develop additional applied science and extension/outreach capabilities in capture fisheries and aquaculture, and to better organize long-term, strategic and medium-term implementation plans and regular impact assessments of an expanded USAID portfolio in capture fisheries and aquaculture.

USAID needs to bridge the "digital divide" to develop solutions to fisheries and aquaculture issues in developing countries.

The Internet, supported via cable and satellite, can be an outreach pipeline, not only for spatial analyses, but also for moving more general information on fisheries and aquaculture to developing countries. There is a vast storehouse of fisheries and aquaculture information in the form of on-line technical reports and publications from US government and state agencies, NGOs, universities, professional organizations, and industry that could be "piped" abroad via the Internet. In many cases, simply raising awareness of the availability and location of the material would suffice as a useful intervention. Distance learning via the Internet, TV and radio is increasingly being used for training in developing countries. Even though, the US possesses the means to be effective in these media, training on fisheries and aquaculture using these media is not yet common. Solutions to the problems of fisheries and aquaculture share a fundamental need of basic comparative information on causes and pathways. Nearly all of the issues are not isolated; rather they are shared by neighbouring communities and countries, or topically right around the globe. Furthermore, many of them already have been experienced and treated in many areas of the world, prominently in the USA.

A fundamental problem is that the applied science that has been employed and experience gained in surmounting the issues is not yet readily available to the developing world at acceptable costs.

Therefore, an important advancement of USAID could be an initiative to help bridge the "digital divide" by increasing its emphasis on identifying, compiling and packaging solutions to fisheries and aquaculture issues of developing countries, and by broadening USAID involvement with many organizations specialized in technical, social and economic solutions in fisheries and aquaculture for which the USA has a comparative advantage—especially by taking advantage of the latitude within USAID for developing a broad variety of initiatives that include the US federal government and state agencies, commercial firms, NGO's, Land/Sea Grant universities, regional aquaculture centers, and international organizations. Several kinds of USAID initiatives are required to:

- Synthesize, package, and deliver applied research information by enhancing extension systems to move information to developing countries and to disseminate it in an effective manner;
- Assess the potential of distance learning to significantly improve technical and managerial competences in fisheries and aquaculture in developing countries;
- Facilitate information flow via choosing methods of information delivery that are appropriate for different audiences in different regions (increasing Internet access, radio, CDs, DVDs, videos, etc.);
- Sponsor collaborative in-country research between US professionals

and applied fisheries and aquaculture professionals in developing nations on problems that are indigenous to the country or region.

DEVELOPMENT OF THE AMAZON AQUAFORUM

We applaud the Pond Dynamics/Aquaculture CRSP in its development of the Amazon Aquaforum, an Internet-based information exchange aquaculture network in South America which supported development of a variety of technologically appropriate information delivery methods. The CRSP also maintains an invaluable, comprehensive, standardized database of information collected from throughout the world. The CRSP developed a web-based resource for small and medium-scale farmers in Latin America called the Web-based Information Delivery System for Tilapia (WIDeST), a decision-making tool that enables users to gain access to useful resources when deciding on appropriate, site specific methods and aquaculture practices.

USAID should prioritize the improved management of coastal marine and inland fisheries by providing technical assistance to evolve innovative fisheries management schemes in developing countries including but not limited to: property rights, co-management, and the use of marine protected areas; plus assist in the development of more accurate and reliable fisheries and aquaculture data reporting systems.

USAID needs to invest additional resources to assist in achieving the long-term goal of sustaining the world's invaluable marine and inland capture fisheries, which are disproportionately located in developing countries. Additional investments in capture fisheries could positively impact and better leverage USAID's current and planned investments in coastal area management and aquaculture.

Additional assistance is needed to engage users and government institutions to bolster management and governance structures to address issues of overcapacity, access, marine tenure, critical habitats and nurseries, and a priority range of social issues such as gender relations. Investigations are needed into conservation engineering (innovative gears and management); roles of reserves, protected and conservation areas; plus investigations regarding how best to protect freshwater flows to estuaries (and rivers to lakes)—especially the timing, volume, quality and pulsing of freshwater flows to critical estuaries.

In addition, the USAID could play a major role in analyzing and promoting effective fisheries management, government policies, reforms, market/trade policies, and reductions of subsidies. In comparison to the long-standing emphasis on technical investments in fisheries for stock assessments, etc., comparably little attention has been paid to the users of fishery resources. This is despite the fact that the observed successes, failures, and constraints experienced in marine and inland capture fisheries management are social and economic in nature. Another problem is that fishery researchers often have a low status and income in a given national context, have limited facilities and resources, few opportunities for in-service training, and have limited access to outside scientific research information. We recommend that the USAID expand its investments in the community-based management of marine protected areas, property rights, and overall investments in social ecology and ecosystems-based management of fisheries.

Illegal, unreported and unregulated fishing is found in all capture fisheries, irrespective of the location, species targeted, fishing gears employed or level and intensity of exploitation. USAID should work with the reporting countries to improve fishery statistics, primarily to meet national needs with regard to food security and fisheries management. Unlike capture fisheries, the separate monitoring of aquaculture is relatively new in most countries, and often there are less well-established systems of data collection as compared to capture fisheries.

USAID needs to substantially increase its support to develop more comprehensive, sustainable, ecologically and socially compatible, environmentally-friendly and economically viable aquaculture systems in developing countries that have the long-term goals of poverty alleviation and food security.

To meet global demands for fisheries products, aquaculture will continue to grow at a rapid pace over the next 10 years; then, its rate of growth will slow until the considerable environmental constraints it faces are solved. Aquaculture faces a number of important problems, including access to appropriate technologies, lack of comprehensive, inter-sectoral planning, and a lack of financial resources for the poor, information on its environmental and social impacts and diseases. The priority areas for further applied research support include:

- Land, water and feed/nutrient use in aquaculture in comparison with other animal protein production systems;
- Sustainable intensification and non-consumptive water use in freshwater aquaculture production;
- Participatory management approaches to the comprehensive development of aquaculture ecosystems as sustainable means of rural development;
- Sustainable coastal aquaculture development, especially technologies that avoid user conflicts;
- Social and economic research to add insights into the adoption of aquaculture by poor rural households;
- Genetically advanced technologies for sustainable stock enhancement and ranching programmes, plus the domestication, selective breeding, and genetic improvement of existing aquaculture species;
- Technologies to solve disease problems and innovative management solutions to improve the health of aquatic animals;

- Development of low cost, non-fish meal based feeds;
- Training in the quality and safety of aquaculture products; and
- Research in making emerging technologies cost-effective, including recirculating systems, and offshore aquaculture systems.

The appropriate role for the USAID in fisheries and aquaculture biotechnology is to support applied research and outreach activities that engage in well-known, conventional genetic improvement techniques, such as selective breeding, etc., as opposed to research support to advanced biotechnologies, such as transgenics. USAID support for The World Fish Center's programme on the genetic improvement of farmed tilapias (GIFT)—and the development of international protocols for product dissemination—is an excellent example of the types of biotechnology investments the merit USAID's future consideration. However, it is also very important that USAID engage nations who have large and active programmes in advanced fisheries/aquaculture biotechnology—such as China—in issues of policies, protocols, environmental, market and other social impacts of transgenics.

USAID should prioritize its assistance to capture fisheries and aquaculture activities that are more integrated, comprehensive, community-based, and use "systems approaches"—such as ecological and integrated farming/fishing systems research and extension approaches—in both rural and urban settings. The current agriculture emphasis of USAID is on plant commodity research, not on a comprehensive, agro/aqua-ecosystems research/extension approach. We urge the USAID to support long-term, applied research and development that makes expanded use of participatory ecological and social science tools to empower community control of fisheries and aquaculture systems; and to better integrate aquaculture and fisheries activities into the comprehensive management of natural and social resources of its missions, target nations and regions.

The kinds of assistance provided by USAID should be based on a combination of: (1) assessed needs and capabilities in developing countries, and (2) the comparative advantages held by the USA in technical expertise, education, communications, business management and commercial products. Assistance focused on US comparative advantages provides a way to get around duplication of effort among competing international organizations, while still fully supporting the Code of Conduct for Responsible Fisheries. The top priority for USAID should be the sustainable development of community-based, integrated farming and fishing systems with the long-term goals of poverty alleviation and food security.

AQUACULTURE AND FISHERIES ACTIVITIES

Aquaculture and fisheries activities are frequently poorly planned and considered; funded separately from activities that greatly impact them; and are generally neglected by all levels of government. Fisheries activities should be planned as a continuum, and comprehensively as part of the planning for integrated natural resources management (water, wastes, agriculture, etc.), and the sustainable development of human communities.

Fisheries and aquaculture information should be linked/coordinated with activities in other sectors (*e.g.*, agriculture, forestry, food processing, distribution, etc.). Fisheries and aquaculture should be planned up-front—not as an afterthought—in all water resource development projects such as reservoir and irrigation projects. Coordination is both justified and essential because of shared issues with other sectors of sustaining biodiversity, maintaining water quality and quantity, the management and development in coastal areas and in river and lake basins, and addressing environmental degradation, mitigation and restoration, and the need to improve governance. USAID is encouraged to prioritize its support to applied research and development activities that articulate well with the natural and social resource and farming systems contexts of a nation/region, and (a) plan for the activities as one part of a comprehensive management strategy for the non-consumptive, multiple uses of water; and (b) use ecosystems-based management approaches that promote the more comprehensive, long-term stewardship of marine and freshwater environments.

We urge the USAID to support the expanded use of participatory tools to empower community controls over fisheries and aquaculture systems; to use innovative co-management methods to sustain local water and coastal resources, and the community-based management of water bodies for fisheries and aquaculture. Fisheries and aquaculture provide foods of very high nutritional value for households.

When resource-poor fishers and farmers combine fisheries, agriculture, aquaculture and the conservation/rehabilitation of natural resources in innovative "ecosystems approaches"' they improve their food supplies, increase their incomes, and become better able to withstand environmental and economic fluctuations; thereby decreasing risks, increasing fishing and farm sustainability, and contributing greatly to rural social and economic development.

We encourage the USAID to incorporate additional, applied social science and micro-economics research into innovative ecosystem-based methods that empower communities to better manage and control fisheries systems, and to develop more environmentally and socially compatible aquaculture systems (protected areas, property rights, innovative co-management approaches, etc.). Most of the modern aquatic resource crises have roots in social issues that are poorly known, such as the "shifting nature of modern survival" in developing nations.

Millions of people do not only fish or farm, rather, they derive income from multiple activities and sources, and in some cases, conduct long distance seasonal migrations between inland farming systems and fisheries systems, and vice versa. USAID is encouraged to support the development of aquaculture systems that are well integrated into existing water resource systems, are virtually non-consumptive of water, and make multiple uses of water.

SMALL BUSINESS DEVELOPMENT

Aquaculture has the potential for small business development but additional technical assistance is needed by business professionals familiar with economics, labour dynamics, opportunity costs, issues of price and volume competition, and other commercial and competitive contexts of other sectors such as agriculture, etc.

The recent activities of the Pond Dynamics/Aquaculture CRSP (PD/A CRSP) are notable in this regard. In 2002, the CRSP conducted regional meetings in Latin America, the Caribbean, Africa, and Asia, plus commissioned a report to explore the current status of aquaculture in Eastern Europe/Central Asia. Participants represented diverse areas of expertise; gender diversity was also a criterion of panel composition.

During the meetings, participants were asked to identify and prioritize constraints to aquaculture development in the region of their expertise. Three central needs-directed programme areas emerged after analyzing the results of the meetings. This movement of the PD/A CRSP to become more of a "system-oriented" network as opposed to a "commodity" collaborative is noteworthy and laudable; and if additional resources were available, these concepts could be developed further.

We cannot emphasize more strongly that poorly-funded, short-term projects with broad, "global" goals will not make lasting impacts on the conservation and sustainable development of fisheries and aquaculture systems in developing countries; and that these approaches do not serve the strategic interests of the United States. Research, education and extension assistance must target the long-term engagement of institutions, and deliver approaches, findings and insights towards these institutional systems and organizations that will continue the long-term engagement with farmers/fishers that can weather the invariable fluctuations in development assistance that will occur over time.

USAID needs to develop comprehensive strategic and implementation plans and regular impact assessments of an expanded capture fisheries and aquaculture portfolio. USAID missions and regions worldwide should include capture fisheries and aquaculture into their strategic plans for the management of natural resources—or they will be incomplete—especially in regards to USAID plans for involvement in the issues of water allocation and quality, and plans for the comprehensive management of marine and inland coastal areas.

USAID needs to sustain a strategic, long-term commitment to fisheries and aquaculture as parts of comprehensive natural, aquatic resource management. Short-term projects should be part of larger, longer-term strategic frameworks for directed action that have adequate accountability to measure strategic progress and impacts. USAID needs to conduct regular, transparent processes that result in the publication of strategic and implementation plans for fisheries and aquaculture over 5 to 10-year time frames. The movement from short term projects to longer term investments in centers of excellence, for example, will require the USAID to develop strategic planning and assessment processes that are much more comprehensive and "living". With more stable investments strategic plans have a shorter "shelf life" and require more frequent review and evaluation in order to make the necessary "mid-course changes". We recommend a process similar to that used by Standing Committees of the U.S. National Research Council where external expert advisors regularly measure progress on investment portfolios.

Within the fisheries sector, USAID should take advantage of shared needs for land and water resources and issues in common to identify sub-sector groupings for directed technical assistance and development activities. We identify two groups that require distinct interventions: (1) offshore (*i.e.*, within the EEZs) and high seas fisheries share common issues of resource and fisheries management that include shared resources, over harvesting, and excess capacity, and (2) coastal and inland fisheries as well as aquaculture to a great extent, may interact both positively and negatively with one another, sometimes competing for space, resources and markets, but they also share many common issues that are external to the fisheries sector, most importantly the environment and poor governance.

DEVELOPMENT STRATEGIES IN FISHERIES

In most countries national economic and social policies influence the formulation of policies for fisheries and aquaculture. Public strategies adopted for development are thus guided by policy objectives which may range from food security to higher incomes or may focus on employment, conservation, the rational use of environmental resources, or on an increase in foreign exchange earnings.

Historically, public strategies for fisheries management relied on centralised, public-funded, fisheries administrations to be implemented. These administrations would assume a range of different functions, including the formulation and enforcement of regulations, undertaking scientific research on which to base decisions on levels of harvesting, and allocating and administering regulations that were formulated. But, fishery administrations in many poor, developing countries often still do not have the means to effectively manage their fisheries.

Thus, the need for innovative strategies, involving a greater participation by stakeholders, is increasingly being recognised.

Direct payments by the fisheries sector for management services in a few developed countries has led to closer scrutiny of the services offered by fisheries authorities, with a view to reducing them to those that are considered to properly be within the public domain, such as MCS and quality control. The fisheries administrations of many of the poorest developing countries are not able to execute effectively the tasks they were established to undertake because they do not have the means to do so. The development and maintenance of skill levels of personnel in fisheries administrations through training is an essential element in achieving sustainable fisheries development.

FRAMEWORKS

The cornerstone for the survival and growth of fishery dependent communities and regions is the economic viability of the fishing activity. Most public sector fishery policies include strategies aiming to establish institutions - or public sector frameworks - that will ensure that communities are helped in maintaining and improving the viability of their fishing/aquaculture activities. A major role of these institutions is to assure that viability is improved without causing negative externalities of any kind.

Background

Historically, the public institutions primarily and directly concerned with fisheries and aquaculture in rich, industrialized countries were centralized in a public fishery administration. The administration would aim: (i) to monitor fishing effort and enforce fishing regulations (MCS); (ii) to monitor fish stocks and advice on appropriate levels of exploitation of major commercial stocks; (iii) to report on the volume and value of production in the sector; (iv) to inform and instruct fishers on new technology through extension services; (v) to develop new fishing and fish processing technologies through public research centres; and, sometimes, (vi) to run offices dedicated to promoting trade and/or consumption of fish products. In addition, within government, the fishery administration functioned as a spokesperson for the sector vis-à-vis other government departments and other sectors of the economy. With growing economies, increased mobility and expanding trade the outside demands on the sector stretch continuously, placing considerable demands on the fishery administration. However, with the advent of policies linked to open market economies, fishery administrations in industrialized countries came under pressure to reduce in size. In addition to a general downsizing, this led fishery administrations to request that users pay for the services that those same administrations provided.

Development

The success of policies derived from the open market economy varies. A main reason for, at best, partial success is that some functions can be reduced in size or eliminated; others not. For example, while arguments can be found for discontinuing technology development and extension services in rich industrial countries, even these countries require that publicly controlled MCS services, as well as the monitoring of wild aquatic resources, continue.

In several countries, public funding for technology development in capture fisheries has been reduced. There are two major reasons for this. First of all, resources are being fully used and where effort is not controlled, improved technology leads to overexploitation, which in turn often causes a decline in livelihoods for the majority of the stakeholders. This of course reduces the urgency of government to participate in developing new technology. A second reason is that the know-how and resources needed for technology development are becoming more and more expensive and sophisticated. In fact, the public sector cannot afford the staff and infrastructure needed to compete in this field with vessel, engine and fishing gear manufacturers. Similarly, as the manufacturers of gears and equipment introduce most new technology, the need for extension services is falling.

Other functions - such as quality control of fish products and MCS of fishing effort - cannot be so easily abolished or reduced. For the well-being of the fishery sector it is essential that MCS remain under the control of the public administration, although it is one of the most costly services provided to fishers. And the need for MCS persists even where entry controls have been introduced.

Where quotas are used it is still necessary for some authority to monitor that quota holders do not exceed their share. This is because quotas do not generally correspond with an exclusive economic or fishing zone. However, successful entry limit schemes often lead to the generation of economic rents for those who remain in the fishery. Thus the industry should be able to contribute to the payment of MCS and resource monitoring services.

Another equally important function - that does not lend itself easily to private sector execution - is the monitoring of the state of wild stocks. Although it remains a costly undertaking for most fishery administrations, it must be continued and, if possible, improved.

During the second half of the last decade, governments in some developing countries had the time and resources to establish full-fledged fishery administrations. But only in the very large countries did these institutions become well established and fully supported through the public budget. In smaller countries with important fisheries such administrations often received significant support in the form of staff, equipment and funding through international aid. But most of the fishery administrations had not become economically self-sustained by the time aid ceased and outside forces constrained governments to adopt policies inspired by the open market economy ideal. For many countries this put an effective stop to efforts aimed at developing appropriate fishery administrations.

Today, in many of the poorest developing countries the fishery administrations are not able to effectively execute the tasks they originally were intended to undertake because they are not provided the means needed to do so. Unfortunately this also reflects negatively on the administrations' capability to act as a spokesperson for the sector vis-à-vis other government departments and other sectors of the economy.

As can be expected - given the costs and infrastructure involved - fishery administrations in developing countries have particular difficulties in monitoring the status of resources and in monitoring and enforcing capture fishery regulations. Also, the resources dedicated by governments to technology development in fisheries are insignificant. The exceptions are large countries - mostly Asian - with large fishery sectors (*e.g.* China, India, Thailand, Indonesia). Staff assigned to undertake extension services are few and have difficulties in performing their tasks for lack of funds and training.In small and poor developing economies - especially in Africa - the situation is indeed critical. As population grows and the fishing pressure mounts on a finite resource, the need for effective resource monitoring and MCS becomes acute. This is true even as more use is made of a property rights-based approach to fishery management. Furthermore, at this time it is unrealistic for most of these countries to consider assuming publicly executed technology research and development programmes in support of either fisheries or aquaculture.

Effective and easy communications between fishers and the administration is a major problem in poor countries. Where such communications are deficient, it is a severe obstacle for development of the sector as there is a growing recognition that the fishery and aquaculture sector, like others, must be governed in part by, and above all with the consent of, the stakeholders. Transparent and frequent communications with stakeholders is a precondition if a consensus is to be achieved on a realistic approach to the management of the sector. For such communication to take place it is becoming increasingly important that fishers are organized and that the management of their associations be prepared to interact with the representatives fishing communities.

SMALL-SCALE FISHERIES DEVELOPMENT

The importance of small-scale fisheries as a major source of animal protein, income and employment cannot be overemphasized. It accounts for a substantial part of the world fish production - at least 40 per cent - and is particularly important in Africa and Asia. Small-scale fisheries provide employment for millions of fishers directly engaged in fishing activities, including rural aquaculture, and for millions more working in fisheries-related activities such as fish processing and marketing, boat building and net making. Including family members, hundreds of millions of rural people in developing countries depend on fisheries for their livelihood. In addition, in many least developed countries of Africa and Asia, fish accounts for more than 50 per cent of the total animal protein intake. In almost all these countries small-scale fisheries provide over three-quarters of the domestic fish supply. In southeast Asia, possibly a billion people rely predominantly on fish for animal protein. A fundamental problem of most small-fishing families around the developing world is their comparatively low standard of living and, frequent poverty despite decades of remarkable overall fisheries development and national economic growth.

Sector characteristics

Small-scale fisheries are scattered along riverbanks, estuaries, seashores and around lakes, often away from mainstream economic, social and political developments. In some areas, their development continues to be severely hampered by the lack of infrastructure such as roads and communication facilities.

They are extremely exposed to natural calamities like floods, inundation, sea erosion and storms and, in many instances, fishers are particularly affected by water-related parasitic diseases like bilharziasis, river blindness, filariasis or malaria. As a consequence of the diversification and seasonality of their activities, many small-scale fishers show high geographic mobility and may migrate over hundreds of miles.

Most small-scale fisheries are exploited under some sort of open access regime, sometimes enforced by modern governments, even though traditionally social mechanisms may have existed to restrict such access. In combination with increasing fish demand and commercialization this has led to excess fishing capacity, resource depletion, waste of economic and human resources, and poor returns on development efforts.

The per capita investment in the means of production is generally low but has increased significantly in many small-scale fisheries during the past two and three decades because of motorization and investments into more efficient fishing gear.

In many instances, productivity has not increased commensurate with the higher investment costs because nearshore and inland fishery resources have become fully, or even, overexploited. The greater harvesting cost could only be covered because of real price increases of fish and fishery products, often fuelled by export markets.

Many artisanal fishing families continue to rank among the most disadvantaged groups of the population, together with landless agricultural labourers and marginal farmers. The incidence of absolute poverty is probably high but there are few studies on the extent of poverty in fishing communities. Availability and access to social services in fishing communities is often below average resulting in low educational attainment and poor health conditions.

Major development constraints

Natural resources such as arable land, water, forests, and fisheries are becoming increasingly scarce. Competition for these finite resources is increasing, as is the vulnerability to the negative effects of other human activities such as pollution and land-reclamation. Inshore stocks are heavily exploited or depleted. In the present management context, the addition of technology provides only temporary relief and increases problems in the medium and long term. Population keeps growing fast in many developing regions. Conflicts between artisanal and industrial fisheries for resources and on the market are increasingly frequent and may jeopardize development efforts.

The major development constraints relate, on one hand, to inadequate access or utilization of social infrastructure including schools and health services and, on the other hand, inadequate institutional arrangements for the management of coastal and inland fishery resources and the protection and conservation of critical fish habitats such as coral reefs, mangroves and seagrass beds. Their de facto open-access nature makes aquatic resources an attractive livelihood option for rural people.

However this advantage constitutes also a threat to sustainability in he absence of access regulation with particularly negative impacts on those with exclusive dependence on aquatic resources. Groups of fishers often have limited alternative livelihood options and this makes them particularly vulnerable to changes in the condition of and access to the aquatic resources on which they depend. From their point of view, greater control of resource use is advantageous as long as themselves are included among those with access rights.

Major development opportunities

In social, economic and ecological terms, small-scale fishers are low-cost producers. Compared to large-scale fisheries, they consume much less energy and require less capital per ton of product and are often less ecologically destructive. They are more flexible and efficient in the exploitation of near-shore resources, provide more employment opportunities, require less infrastructure and are often less subsidized than industrial fisheries. Decentralized and highly integrated in the local economy, they contribute to rural development and slow down the population drift to urban centres. They are a unique source of livelihood for poor sections of the population and contribute effectively, and sometime very significantly, to food security and nutritional balance in rural areas. Even when resources are fished close to the maximum possible it is feasible to reduce fishing costs, improve market value through better processing, value-adding, and marketing. It is also possible to develop complementary or alternative employment through other productive activities, such as aquaculture, agriculture, animal husbandry, handicrafts.

Development requirements

The sustainable development of small-scale fisheries is desirable for a balanced social, economic and regional development in coastal and rural areas. It would require, inter alia, increased political and economic support; more directly applied socio-economic and policy research; more favourable fishery development policies and strategies; integration of fisheries into rural development and coastal areas management; better identification and allocation of resource rights; stronger protection of reserved fishing areas from intrusion of large scale fisheries (which will require elimination of overcapacity in these fisheries); enhancement of competitiveness in using resources; adoption of decentralised, participative management processes; facilitating the development of strong local institutions as a vehicle for empowerment and decentralisation; control of fishing capacity and creation of alternative employment; promotion of technological progress, to reduce negative impacts on the environment and improve product quality; better access to credits and inputs, markets and services.

TRAINING AND FISHERIES DEVELOPMENT

UNCED and its Agenda 21 recognized that the lack or shortage of national capacity is one of the main impediments to the sustainable development of fisheries. Capacity-building is needed to improve inter alia the qualifications of manpower (*e.g.* fishers as well as scientific advisers or fishery administrators), the technology and the institutions (including laws and regulations, organizations and processes). Training and extension are two important elements of capacity-building.

Training for the fishery sector is no less a continuous process than in any other sectors, nor is the basic need for training any different. This applies to developed and developing countries alike. However, specific requirements differ greatly within the fishery sector and the opportunities as well as standards of training are often not yet consistent with the actual needs, particularly in many developing countries.

This is partly due to the fact that the scope of training needs for the wide range of disciplines in the fisheries sector is not always clearly defined. A contributory factor in this respect is inadequate manpower planning programmes and in many cases the numbers of persons involved in the post-harvest and culture sectors are not known, and data on those employed in capture fisheries could still be improved.

Furthermore, for those employed in the fisheries sector, standards of education, training and certification are generally set by administrations or by international organizations which may not have executive responsibilities for fisheries. This means that a high degree of cooperation between government ministries or departments, and between international organizations is required in order to address fisheries-related subjects and such cooperation is not readily achieved in all cases. Consequently, there is still concern that fishery education and training needs might not be adequately catered for in some countries.

The need to address problems associated with education and training in fisheries is not new. The FAO Committee on Fisheries has always given a high priority to training, as demonstrated by the establishment of a Sub-Committee on Training and Education at its Second Session in April 1967, the priorities at that time being reflected in the Fisheries Department's Regular and Field Programmes.

However, the development of technology in subsequent years led to greater demand for applied courses at technical and professional levels, to develop self-reliance in fisheries management and development. This was particularly so in the mid-1970s when the discussions at the United Nations Conference on the Law of the Sea resulted in global acceptance in 1982 of the coastal States' authority to manage fisheries within their extended economic zones (EEZs).

The FAO World Conference on Fisheries Management and Development, Rome, 1984, acknowledged that training was, and would continue to be, an essential part of all Programmes of Action and a prerequisite for further development, in order to sustain the fishing industry. It noted that general and applied education programmes would have to be improved in many cases in order to obtain maximum benefits from training.

Principles and guidelines

The principles and guidelines with respected to training which were embodied in the Strategy for Fisheries Management and development adopted by the Conference have also been applied to the Regular and Field Programmes on a national, as well as regional, basis:

- Training programmes should be elaborated within the context of overall national fisheries development plans and management policies.
- Training programmes should be based upon clearly defined needs and realistic assessments of existing trained or experienced manpower and current technology. Training should be categorized as to subject and target recipients. Governments should also determine which programmes might be undertaken using local resources and which ones need regional and extraregional expertise.
- While the major responsibility for basic training remains at the national level, external financial and technical assistance is often required to strengthen national training capacities and to provide supplementary training, particularly for higher-level personnel and in specialized skills.

- Because of the increased emphasis on policy formulation and the design and implementation of management schemes, training programmes should be organized for high-level administrators responsible for the fisheries sector. Training of mid-level personnel is equally important for implementation of development plans. Facilities should be maintained for continuous professional training.
- Efforts should be made to establish post-graduate courses and to develop improved training materials. To that effect, close cooperation at national level between fisheries research and academic institutions active in marine sciences and at international level among the appropriate UN organizations should be encouraged.
- Particular attention should be given to the training of extension staff and training specialists and of senior and intermediate-level key staff of various disciplines, who can contribute to training and assist in bringing about a multiplier effect.
- Improved training techniques and methodologies need to be developed and special training equipment and simple materials should be prepared to meet the needs of illiterate and semi-literate people. In this respect, efforts to raise the general educational level in fishing communities are important. Institutional and on-the-job training of both deep-sea fishing crews and artisanal fishers should be enhanced.The use of fishery cooperatives for training purposes should also be encouraged.
- Particular attention should be paid to the design and monitoring of on-the-job training programmes for selected fishing communities. Attention should be given to the training of local fishers and fisheries administrators in basic resource management, in environmental protection, in the operation and management of fishers' organizations, and in activities associated with social development.
- Governments should endeavour to provide incentives to ensure that those trained in fisheries are effectively deployed and retained within the fisheries sector.
- Training programmes should be regularly reviewed to ensure their effectiveness and relevance to needs.
- The transfer of technology should be promoted through pilot projects and assistance in building necessary infrastructure, product development and marketing.
- Encouragement must be given to the education of consumers to that they can take fuller advantage of the nutritional benefits of fishery products, particularly in regions where there is no tradition of consumption. Such training should include instruction in ways of preparing fish products.

THE COLLAPSE OF MANY OF THE WORLDS FISHERIES

LEGISLATIVE TRENDS IN CAPTURE FISHERIES

Over-fishing has led to the collapse of many of the worlds fisheries, in many instances because of the existence of open access regimes where the fisher chooses where and when to fish and how much fish to take. Such regimes were justified largely on the erroneous belief that the oceans and other water bodies hold infinite fish resources. In addition, the oceans, in particular the high seas, were *res communis* so that everyone had a right to fish and no one had ownership over the resource or the right to limit access. While many large areas of oceans are now subject to the jurisdiction of coastal states, and high seas fishing is subject to rights and obligations under the 1982 UN Convention, one may be surprised to find that many states have only recently begun to regulate domestic fishing or to consider stringent control over fishing.

In many Southeast Asian states located to the West of the South China Sea, for example, an individual has an unquestionable right to take to the sea to fish, provided that he or she complies with whatever requirements or regulation is in place. In Thailand, an act such as registration of a fishing vessel which is a prerequisite for entering the coastal marine fishery is a mere formality. In this situation, the fisheries management authority in general considers its role to be one in the service or support of fishers or for the promotion and development of fishing. In Tonga, fishing, particularly small-scale fishing, is considered a livelihood, and no licences have been issued for medium- to large-scale commercial fishing until recently. In many cases government regulation is seen as an unnecessary hindrance to the right to fish. However, the trend today is that governments are realizing that in order to ensure the long-term sustainability of fishery resources, the once open access regime for fishing can no longer continue. To this end, Thailand and Tonga, among others, are considering legislation which could make the right to fish subject to significant controls. Given that it could still be a matter of opinion whether anyone has a right to fish in certain jurisdictions, governments like Iceland enacted fisheries legislation that attempts to clarify the issue as follows:

"Marine resources that are found in Iceland waters and are utilized are the common property of the Icelandic nation. The purpose of this legislation is to ensure the preservation of and sensible utilization of these resources thereby guaranteeing full employment and stable settlement of the country. The issuing of fishing permits, in accordance with this legislation, does not constitute any claims to ownership or irrevocable claims by individual parties over fishing rights".

Senegal makes a similar statement in its 1998 fisheries legislation. Even countries that have never before regulated fisheries, such as Ethiopia, are considering new policies and legislation to properly manage the significant increase in fishery activities and to prevent over-exploitation. As a land-locked country, Ethiopia's fishery activities occur entirely in inland water bodies such as rivers, lakes and reservoirs. The proposed fisheries laws for Ethiopia set out the basic framework for fisheries management and contain many of the principles referred to above, including the precautionary approach. Other jurisdictions have focused more on asserting the right of the government to manage fisheries resources. Thus Namibia recently stated in legislation that the management, protection and utilization of marine resources in Namibia and Namibian waters shall be subject to its Marine Resources Act of 2001. Cameroon and the Marshall Islands set out the right of the government to manage and control the fisheries resources in stronger terms. A similar approach, in the context of inland fisheries, can be found in Malawi's legislation.

TOWARDS THE USE OF PROPERTY RIGHTS

Related to the trend away from open access to limited access regimes is the move towards creating property rights in fisheries resources and the allocation of such rights. The move from an open access to a limited access regime is in essence a move from one form of administration of property rights to another.

A property rights regime can be a state property regime, private property regime or common (collective) property regime. The most universal is the state property regime, although most states do not commonly depend on "property" or know and administer it as such, but rather on the distribution of fishing rights on the basis of jurisdiction or sovereignty. In such a regime, the state is the custodian of the fishery resource and can decide to leave the resource to free use (open access), can choose to exploit the resource directly through its own agencies or can allot rights to citizens to exploit the resource. The most widely used and practical system of administering the last variant is that the state grants licences to individuals or groups to fish, and in this way controls access to the resources. At the other end of the scale is the common property regime whereby a local community instead of an individual holds exclusive rights to harvest fish in a certain geographical area.

The developments in Eastern Europe and the former Soviet Union illustrate the effects of the different property regimes. Before the shifts to market economies, fisheries were maintained at an almost constant level by centralized economic plans. However, during the transition period, the fisheries operated under an open access system which stimulated competition but at the same time increased the risk of over-exploitation of the stocks. Therefore, new fishing regulations have been established that set out fish licensing and catch quota regimes, to minimize the risk of collapse. For example, Lithuania enacted a Law on Fisheries in 2000.

Under the private property regime, normally the authorized user who has received a licence has a personal right to fish or harvest and such right is renewed regularly. More recently, licensing schemes have begun to permit the authorized user to sell or lease the right to fish. Such a property regime requires the setting of the total allowable catch (TAC) for a specific fishery, and establishes systems for allocation of the TAC, the transferability and leasability of the rights and the manner in which such rights can be enforced. Iceland has institutionalized the individual quota system in legislation, as has New Zealand. In Africa, Angola and Mozambique have legislative provisions that will enable the use of a property rights system in the future.

FROM DEVELOPMENT TO SUSTAINABLE UTILIZATION

Fisheries management prior to Rio generally emphasized optimum utilization of fishery resources. In respect of marine capture fisheries, this perspective is reflected in the 1982 UN Convention, which calls on states to optimally utilize fisheries resources. Many fisheries management authorities at that time focused on building capacity or encouraging entry into the fisheries industry to increase fish production for individual profit, revenue generation or domestic consumption. Fisheries management was also an exclusive mandate, often devoid of considerations of the impact of fishing or fisheries activities on associated matters such as the environment. Conversely, other industries did not consider the impact of their activities on fisheries.

There has been a gradual shift in focus to a more "rounded" approach to fisheries management - one that not only ensures exploitation of the resource for economic gain, but also ensures that the resource is maintained at biologically, environmentally and economically sustainable levels. The shift in focus which has been forged in international fora is slowly becoming entrenched in domestic policy and eventually has found its way into legislation. A few examples demonstrate this.

Whereas in the past the statement of purpose in fisheries legislation might refer to allocation of a quota levy, the development of fishery or the management of fishery, in recent times one sees an inclusion of a statement of principles and policy relating to sustainable utilization of fishery resources, as in Namibia, Nauru, New Zealand, Papua New Guinea and South Africa. For example, the Marine Resources Act of Namibia has as its purpose to "provide for the conservation of the marine ecosystem and the responsible utilization, conservation, protection and promotion of marine resources on a sustainable basis; for that purpose to provide for the exercise of control over marine resources; and to provide for matters connected therewith". The South Africa Marine Living Resources Act of 1998 has a similar provision.

While preambular provisions of a fisheries act are usually not enforceable, these statements of purpose are given effect in the law's operational provisions, *i.e.*, in the mechanisms it establishes. More significantly, the strong reference in such laws to conservation of the ecosystem and to long-term sustainable utilization is important for it stamps them as very much of the new era of fisheries management.

Similarly, management action was, in the past, based primarily on scientific (biological) information directly affecting the species or fish stock in question. Little or no consideration was given to the effect of fishing on species associated with or dependent on the target species or on the aquatic environment. National legislation now calls for the effects of fishing on non-target or associated species or the aquatic environment to be considered in determining the types of management measures that should apply in a fishery, and to what extent. Legislation that demonstrates this trend can be found in New Zealand and South Africa.

Laws of other jurisdictions may not necessarily refer to environmental protection, as in New Zealand, or allow for environmental impact assessments, as in South Africa. Nevertheless, other management actions required under fisheries laws may have the effect of protecting the environment. Sustainable use of the fish stocks and the protection of the environment are now key elements in the recently developed legislation and institutions of Albania, Hungary, Lithuania and Romania.

A related and interesting feature is an increase in references to the precautionary approach or principle. Such references are viewed as an effort by states to give effect to international fisheries instruments or agreements, particularly principle 15 of the Rio Declaration, the UN Fish Stocks Agreement and the Code of Conduct. Even if the precautionary approach is only mentioned in a preambular provision or a broad policy statement and may at best be used only as an aid to interpretation, its significance should not be underestimated, for it "represents a major change in the traditional approach of fisheries management, which until recently has tended to react to management problems only after they reached crisis levels".

IMPROVED ENFORCEMENT

As countries seek to ensure that only sustainable levels of fishing activity are allowed in zones under national jurisdiction, several innovative mechanisms and legislative approaches to fisheries monitoring, control and surveillance have emerged, in particular to curb illegal, unreported and unregulated (IUU) fishing.

Vessel Monitoring Systems

The use of satellite-based vessel monitoring systems (VMS) is a recent development in fisheries monitoring control and surveillance (MCS). VMS ensures that fishing vessels provide reports in real time. VMS can be seen as a direct response to IUU fishing, in particular fishing that is unreported due to problems with radio reporting systems and other conventional means of reporting vessel positions.

At this time, VMS is focused on position reporting, although other VMS information, namely, sighting reports, catch reports, notifications (entry/exit into the exclusive economic zone, port entry, etc.) and analyses, can also be generated by VMS. VMS is currently considered a complementary tool to conventional MCS tools such as sea and air reconnaissance. VMS is in use or in various stages of trial and implementation in many countries and by regional fishery bodies. In the case of one regional fishery body, the Forum Fishing Agency (FFA), the member countries that are mostly developing countries sought to overcome their individual limited resources for MCS by establishing a regional VMS. The issues in the implementation of VMS which are dealt with in legislation are: requiring the installation of VMS components, namely automatic location communicators or vessel tracking units; protecting VMS components; ensuring confidentiality of VMS information; and using VMS information in fisheries enforcement in courts, particularly the admissibility of VMS information such as vessel positions. Recent legislation on VMS in Europe, North America, Southern Africa and the South Pacific has sought to address these issues.

"Long-arm" Approach to Enforcement

Another mechanism that has emerged in fisheries law enforcement is a provision in national fisheries legislation commonly referred to as the "Lacey Clause", from the Lacey Act of the United States. The provision extends the arm of the law, basically by making it unlawful to import fish that has been taken contrary to the laws of another country. A common example of violation of the laws of another state is the taking of fish without a licence where such licence is required by that states' fisheries legislation.

A typical Lacey Act provision states that anyone who lands, imports, exports, transports, sells, receives, acquires or purchases any fish taken, transported or sold contrary to the law of another state is guilty of an offence and liable to a fine. Such a clause was first adopted in the FFA region by Papua New Guinea in 1994, followed by Nauru in 1997 and Solomon Islands in 1998. New Zealand has introduced in a recent amendment to its principal fisheries legislation a Lacey Act-type clause, which prohibits its nationals from fishing in another jurisdiction in contravention of that jurisdiction's laws. An interesting aspect of the New Zealand legislation is that the prerequisite element of bringing fish into the country is not necessary, although such extra-territoriality applies only to New Zealand nationals and vessels.

Alternatives to Criminal Proceedings

Despite innovative efforts such as enactment of Lacey Act clauses, enforcement of fisheries provisions through criminal laws and procedures has a number of drawbacks. There is a high standard of proof, and there may be difficulties in using evidence generated by VMS due to the hearsay rule. In many jurisdictions, extended delays plague the criminal law system. One solution, now applied in the United States, the FFA region and many civil law countries, is the adoption of civil and administrative processes and penalties for dealing with fisheries offences. This approach presents the advantages of expedited proceedings, lower standards of proof, possibilities for negotiated settlements and hearings which do not necessarily follow strict rules of evidence. Civil penalty schemes for fisheries violations treat certain violations of fisheries laws as civil wrongs penalized by civil penalties, while the right of the offender to decide that he or she be tried under the normal judicial process is preserved. While only a few countries have adopted administrative proceedings to deal with fisheries offences, many states in the Caribbean, Indian Ocean and South Pacific regions have adopted a system of compounding of offences in order to deal swiftly with fisheries violations.

The main element of such a scheme is that instead of resort to traditional court proceedings, the person in whom powers to compound offences is vested (usually the Minister responsible for fisheries or the chief executive officer in the fisheries administration) decides to accept sums of money - usually not more than the maximum of fines allowed - from the offender if it is believed that an offence has been committed. Other requirements in more recent legislation are that offences may be compounded only with the consent of the alleged offender and that the Minister or chief executive officer may be empowered to release any article seized in relation to the offence. The offender retains the right to have the matter against him or her heard in normal judicial fora.

2

The Sustainability of World Fisheries

Concerns about the future sustainability of global fisheries have prompted responses at both the national and international/regional level.

INTRODUCTION

The concept of sustainability has been brought to the center of socio-economic and environmental debate after the well-known definition of sustainable development by World Commission on Environment and Development. However, since then the notion of sustainability has generated tremendous interest and an avalanche of publications, even though it has never found an agreed definition of sustainability itself.

Most of sustainability definitions originate from the relationship between humans and natural resources system. Wimberly states that ˉto be sustainable is to provide for food, fibre and other natural and social resources needed for the survival of a group—such as a national or international society, an economic sector, or residential category—and to provide in a manner that maintains the essential resources for present and future generations. Norton argues that ˉsustainability is a relationship between dynamic human economic systems and larger, dynamic, but normally slower changing ecological systems, such that human life can continue indefinitely, human individuals can flourish, and human cultures can develop—but also a relationship in which the effects of human activities remain within bounds so as not to destroy the health and integrity of self-organizing systems that provide the environmental context for these activities.

Constanza, in line with the above definitions, defines sustainability in systems properties, stressing that ˉsustainability…implies the system's ability to maintain its structure (organization) and function (vigour) over time in the face of external stress (resilience). Moreover, more on economic dimension, Solow argues that the system is to be said sustainable as long as the total capital of the system is equal or greater in every next generation. Constanza and Daly mentioned that sustainability only occurs when there is no decline in natural capital.

From several definitions of sustainability, there is one common component in all of them. There is something about maintenance, sustenance, continuity of a certain resource, system, condition, relationship between components of the system, in keeping something in a certain level, and of avoiding decline.

In fishery sciences, the sustainability has long been in discussions and debates on the concept of sustainable yield. Sustainability in fishery has been determined predominantly by the level of sustainable catch which Charles called as the conservation paradigm of sustainable fisheries. Furthermore, it was elaborated that the conservation paradigm in sustainable fisheries makes the definition of sustainability as one of long-term conservation, so that any activity is judged ¯sustainable if it protects the fish stocks without considering much on human-oriented fishery objectives. The icon of MSY (maximum sustainable yield) is one of the very well-known parameter that has been emphasized by the conservation paradigm.

Another perspective, called the rationalization paradigm, has challenged the dominancy of conservation paradigm by inserting the concept of resources rent, the return to resources owners from the fishery. In this paradigm, fisheries should be run rationally and economically efficient. This rationalization paradigm focuses on the achievement of an economically ¯rational or ¯efficient fishery.

In contrast to the conservation and rationalization paradigms, public policy debates often revolve more around human concerns. This paradigm, called social/community paradigm, views the best means to achieve fishery sustainability is through a complex and systematic analysis on community-based means, capable of controlling harvests, making use of appropriate technology, and promoting long-term resilience and diversity. In this paradigm, sustainability is not achieved by only focusing on the conservation of the fish nor maximizing the economic rents but rather by preserving the way of life in fishing communities.

By the above visions, it is argued that sustainable fisheries should be defined widely. While there is general agreement that resource conservation is necessary for fishery sustainability, the concept of fishery sustainability must involve more options and objectives, including other human concerns instead of rational objectives such as MEY and OSY. Pitcher and Preikshot have also mentioned that conventional stock assessment (conservation paradigm) relates to the ecological, or occasionally the economic rationale paradigm, and yet fisheries in reality are a multi-disciplinary human endeavor that has social and ethical implications, besides policy implications.

Evaluation of sustainability in all of these disciplines is required for objective decision-making. In other words, while the balancing of present and future catches is important, there is more to a healthy future than simply a large fish stock.

It is also important to pay attention to sustaining the process underlying the fishery. Moreover, Charles mentioned that what is sometimes missing from the discussions of fisheries sustainability is the attention to the state of human system. This is where the concept of sustainable development becomes important.

As mentioned in the beginning of this chapter, a well-known definition of sustainable development is the one given by WCED as the development that meets the needs of present generation without compromising the ability of future generations to meet their own needs. Following this definition, there is wide recognition of the need to view sustainability broadly, in an integrated manner that includes ecological, economic, social and institutional aspects of the full system (*i.e.* fishery system).

Ecological sustainability incorporates firstly, the long standing concern for ensuring that harvests are sustainable, by avoiding depletion of the fish stocks. Secondly, ecological sustainability also incorporates the broader concern of maintaining the resource base and related specie at levels that do not foreclose future options.

Socio-economic sustainability focuses on the macro level, that is, on maintaining or enhancing overall long-term socio-economic welfare. This socio-economic welfare is based on the blend of relevant economic and social indicators, focusing essentially on the generation of sustainable net benefits, a reasonable distribution of those benefits among the fishery participants, and maintenance of the system's overall viability within local and global economies. The socio-economic sustainability blends together economic criteria (such as the level of resources rent) and social criteria (such as overall distributional equity), recognizing that these are inseparable at the policy level.

Community sustainability emphasizes the group level, that is focusing on the desirability of sustaining communities as a valuable human system in their own right, more than simple collections of individuals. It is recognized that a community is more than a collection of individuals. Hence, emphasis is on maintaining or enhancing the group welfare of participating and affected communities including economic and socio-cultural welfare, overall cohesiveness, and the long-term health of the human system.

Finally, institutional sustainability involves maintaining suitable financial, administrative, and organizational capability in the long term, as a prerequisite for three components of sustainability described previously. Institutional sustainability refers in particular to the sets of management rules and policy by which fisheries are governed. A key requirement in the pursuit of institutional sustainability is likely to be the manageability and enforceability of resource-use regulations.

As elaborated by Charles, the first three sustainability components can be viewed as the fundamental points of a Sustainable Triangle. The fourth, institutional sustainability, interacts among these potentially affected positively or negatively by any policy measure focused on ecological, socio-economic and/or community sustainability. If each of the components is viewed as crucial to overall sustainability, it follows that sustainable development policy must serve to maintain reasonable levels of each.

In the other words, system sustainability would decline through a policy seeking to increase one element at the expense of excessive reductions in any other. The concept of inseparable elements of sustainability in fisheries could be seen as the sustainability triangle.

ASSESSING THE IMPORTANCE OF SUSTAINABILITY INDICATORS

In this study, a formal methodology called multi-criteria analysis (MCA) is used. According to Mendoza and Prabhu, MCA is a general approach that can be used to analyze complex problems involving multiple criteria, and also have advantages when applied in a complex and stochastic system like fisheries. At least there are three advantages of this method for fishery sustainability assessment.

First, it can deal with mixed sets of data, quantitative or qualitative, including stakeholders' opinions. Fisheries, as a system, are well known to be complex and stochastic so that incomplete information and understandings may exist. In this case, qualitative information from stakeholders, including experts groups, and experiential knowledge have distinct advantages for assessing sustainability indicators of fisheries system.

Secondly, the MCA approach also can be conveniently structured in order to enable a collaborative planning and decision-making environment. This environment provides an opportunity to develop such an accommodation for the involvement and participation of stakeholders in the sustainability assessment process. Finally, the MCA methodology is also still simple, intuitive, and transparent while it has strong technical and theoretical support in its procedure.

In this study, MCA approach was used to (1) generate a set of sustainability indicators of fisheries based on several appropriate references; and (2) assess and evaluate the indicators in terms of their degree of importance and condition with respect to some desired future condition or target. For the first part of analysis, the methods used to identify and monitor sustainability indicators are quite diverse, but they frequently fall within a range from expert driven and top-down to bottom up, and locally defined. In this study, we developed a mixed-method approach suggested by Parkins *et al.* *i.e.* by combining expert-driven fisheries sustainability indicators and then they are confirmed to the local stakeholders in order to generate a ¯local accepted fishery sustainability indicators. It can be presumed that by using this combination approach, the sustainability indicators will be more directly suitable to the community goals and objectives.

Analysis of Indicator Linkages

We have described the methodology of static and individual behaviour of each indicator from the stakeholders' perceived importance and conditions. However, it can be also argued that each indicator seldom affects the dynamics of fishery system and their sustainability individually. They are intricately linked and connected among them and consequently will affect the sustainability directly or indirectly. Their impacts are tied through a web of complex relationship which is difficult to extract on an individual indicator basis. Moreover, Prabhu *et al.* also mentioned that sustainability of a natural system can be compromised due to cross-criterion or cross-indicator interactions. In the other words, it would arguably be meaningful if analysis can be extended into behaviour assessment among indicators using system approach.

According to Mendoza and Prabhu, cross-indicator interactions can be analyzed at different levels depending on the amount of information and knowledge about dynamic interactions between indicators. It can be analyzed using quantitative system dynamics if sufficient information about each indicator can be identified. On the other hand, qualitative analysis and assessment of indicator linkages also can be more suitable in the case of lack of functional relationship between indicators. In this study, we used one qualitative method, that follows an approach with regard to problems of indicator linkages, called cognitive mapping.

Cognitive mapping is included in the soft methodology category and differs from traditional formal methodologies in terms of the type of analysis and results generated. Generally, soft methodology results are descriptive rather than prescriptive. Cognitive mapping is a casually based mapping technique where concepts representing elements of a complex problem are organized and structured using arrow diagram. Arrows represent the connections and relationships among the indicators.

In this study, we defined two variables which Eden and Akermann considered as the essential variables of cognitive mapping method, *i.e.* domain and centrality. Domain is an important factor of cognitive mapping because it reflects the density or the number of indicators directly linked to a particular indicator regardless of direction. That is, higher domain values of an indicator reflect a larger number of indicators directly affecting, or affected by, the indicator.

IMPORTANCE OF SUSTAINABILITY INDICATORS

The first analysis for sustainability indicators is to generate a set of indicators in terms of their importance judged by a group of stakeholders. In this study, we used a set of sustainability indicators which is composed of four variable criteria of sustainability indicators, namely ecological-criterion indicators (5 indicators), economic-criterion indicators (5 indicators), community-criterion indicators (5 indicators) and policy-criterion indicators (3 indicators). A group of stakeholders that consists of three types of stakeholders was involved in this analysis, *i.e.* fishers community (10 persons), fishery-related decision-makers (3 persons) and fish market-related stakeholders (2 persons). By using participatory assessment method, we discussed the indicators set and asked the stakeholders to define their value on indicator importance as well as their perceived target of indicators.

The first result is the acceptance rate of an indicator, that is the rate of stakeholder judgement to the indicator or whether the stakeholder involves the indicator in his/her judgement. From this result, we reveal that catch structure indicator (indicator a.1) has the lowest acceptance rate, *i.e.* out of 15 persons only 5 persons (or about 33 per cent) put their value to this indicator in their analysis. The participation rate for other indicators were calculated more than 33 per cent and maximum at the level of 100 per cent (all of stakeholders involve the indicator to their analysis). From the criterion level, the policy-criterion indicator group has the highest average acceptance rate, *i.e.* all indicators in this criterion were involved by the stakeholders (100 per cent).

The following results show the importance of degree of indicators which are judged using a 7-point scale of values by the stakeholders. It is clear that according to stakeholder values, almost all the indicators are rated moderately to highly important. It can be seen from the average weight value, which is calculated in a range from 5.15 to 6.73, showing that almost of all the indicators are important. Furthermore, from the calculation of relative weights, we can clearly see that some indicators are rated lower than others.

For example, under the ecological sustainability indicators, indicator a.5 (effect of human activities on the marine ecosystem) is rated lower than other indicators under the same variable (ecological sustainability). Similarly, indicator c.11 (demographic indicator) and c.12 (succession rate) under the community sustainability has low relative weight as compared with the other three indicators in the same variable. Moreover, we also reveal that among all of indicators, indicator c.15 (security level) under the community indicators variable has the highest value of importance, while indicator b.10 (fisheries employment/proportion of labour) under the economic sustainability variables has the lowest value of importance degree.

In overall, it is also clear that from the average total weight value, the stakeholders judged the policy and community indicator variables are more important than ecological and economic indicator variables. From this information we can say that rather than ecological and economic objectives, the stakeholders may argue that, policy and community objectives should be stressed more in the general fishery development policy. This is confirmed by the fact that fishery governance in Japan has emphasized on policy and community-based decision making for many years. However, this finding does not mean that ecological and economic objectives are not important. A different picture is also found for the case of stakeholder group interests. The degree of importance in overall is different among group of fishers, policy makers and market-related persons. For example, in fishery group, community and policy sustainability indicator variables are important, while policy makers emphasize on ecology and policy sustainability variables, and market-related persons considered economic sustainability variables as more important variables. These comparative analyses can be used as guides in making decision as to what indicators are important and need to be monitored site-specifically.

From the calculation of standard deviation we also reveal that there is no complete agreement or consensus among stakeholders due to the standard deviation value of more than zero. In the other words, the opinions and judgements are more varied among the stakeholders. Higher standard deviation indicates diverging opinions; the larger the standard deviation, the more varied the opinions. Among fisheries sustainability indicators examined, ecological sustainability indicators has the highest average of standard deviation (1.09), meaning that the judgement or opinions of the stakeholders for the ecological sustainability indicators are divergent and varied among others. On the other hand, due to its average standard deviation value (0.75), it can be said that there is a ¯degree of consensus among the stakeholders when making a judgement for the community sustainability indicators. The economic and policy sustainability indicators have average standard deviation value between those values. According to Mendoza and Prabhu, it would ideally be desirable to have some way of ¯improving the consistency of the judgements (minimum standard deviation).

In addition to estimating the relative weights of indicators, the next part of analysis is to estimate the ¯sustainable state elaborated from the perceived targets or conditions judged by the stakeholders. This analysis is also started by judgements of the stakeholders to score the perceived targets of each indicator followed by the calculation of sustainability index of criteria (SIC). We can see that sustainability index for the ecological indicators is the highest among other sustainability variables (SIC=3.79). It is followed by the economy indicators (SIC=3.57), community indicators (SIC=3.26), and policy indicators (SIC=3.20).

The results of SIC mean that based on their perceived value of condition for each indicator, the stakeholders judged that ecological indicator variable has a relatively best condition (acceptable, above the norm of good operations in the region) compared with other variables. This result can be complementarily discussed with the results of important judgement of indicators. Although, according to the stakeholders, policy and community indicator variables are the most important for fishery system in Yoron Island, they also judged that ecologically and economically, the fishery system in Yoron Island is relatively "sustainable".

Indicator Linkages

As previously mentioned, the indicator of fishery sustainability seldom affect the dynamic of ecosystem or sustainability on its own. Therefore, a holistic assessment is done to identify the relationship among indicators including their collective impact on the fishery sustainability.

In terms of domain, the results show that only 5 from total 18 indicators have a density of at least 5 indicators (*d*-max) directly linked to them *(d=5).* Those indicators are distributed in the economic sustainability variable, 2 indicators (b.7 and b.8), and then the community sustainability variable, 3 indicators (c.11, c.12, c.14). The maximum indicator domain in the other two sustainability variables is 4 indicators *(d=4)* for the ecological sustainability variable and 3 *(d=3)* for the policy sustainability variable.

In terms of centrality score, we also found that indicator c.12 (succession rate) under the community sustainability variable has the highest value (*CS*=27.17), meaning that arguably this indicator can be perceived as the central of sustainability issues for the fishery system in the case of Yoron Island, Japan.

INTERNATIONAL AND REGIONAL RESPONSES

The international response has been manifold and varied. In 1930, the Hague Conference on the Codification of International Law saw the international community recognize national claims to territorial waters. Maritime property rights were subsequently strengthened at the United Nations Conference on the Law of the Sea in Caracas in 1974 when the principle of 200 nautical mile Economic Exclusion Zones (EEZs) and the concept of Extended Fisheries Jurisdiction (EFJ) were advanced.

Although formal adoption (1982) and the entering into legal force (1994) of EEZs and EFJs was delayed somewhat, 150 states had ratified the UNCLOS Convention by late 2006. The more thorny issue of stocks that traversed international maritime boundaries was subsequently addressed through the adoption (1995) of the UN Agreement on Straddling Fish Stocks and Highly Migratory Fish Stocks, 61 states having ratified the Agreement by late 2006. This latter agreement was portentous as, for the first time, the objectives of international fisheries management became sustainable fishing, ecosystem protection, conservation of biodiversity, and the precautionary approach to fisheries management.

FAO too, was active in the realms of policy advice and guidance. The Code of Conduct for Responsible Fisheries and the ensuing Compliance Agreement confirmed that the nation state remains responsible for the activities of national flagged vessels, and seeks to advance management measures that improve the optimal and sustainable use of living aquatic resources. The Code also provided a framework for the subsequent development of four voluntary international plans of action (IPOAs) concerning the conservation and management of sharks, the reduction of incidental catch of sea-birds in long-line fisheries, the prevention of illegal, unreported and unregulated (IUU) fishing, along with the management of fishing capacity.

Wider recognition of the perilous status of many fish stocks was purveyed by the World Summit on Sustainable Development (WSSD) in Johannesburg in 2002. Not only did the Summit signal the importance of the Earth's oceans, seas, islands and coastal regions in sustaining economic prosperity and contributing to global food security, it also advocated the need for multidisciplinary and multisectoral national programmes of coastal and ocean management (reinforced by strengthened regional cooperation and coordination mechanisms) to protect marine resources.

In addition, numerous intergovernmental bodies and non-governmental organizations are presently active in the fisheries/marine conservation arena. Organizations such as the International Whaling Commission and the various Tuna Commissions have mandates to promote the sustainable harvesting of whales and tuna (including tuna-like) species respectively, the 1982 Convention on the Conservation of Antarctic Marine Living Resources (CCAMLR) is charged with establishing managerial control over marine resource stocks in the southern polar region, European fishers are subject to detailed regulations laid down by the EU Common Fisheries Policy, while the Marine Stewardship Council (MSC) certifies fisheries which are harvested in a sustainable manner.

Ultimately, however, major managerial responsibilities devolve to nationally-based institutions.

National Responses

At the national level, fisheries management has attempted to promote the sustainable harvesting of stocks by either blocking or adjusting the incentive to overfish. Incentive-blocking methods seek, by curtailing the incentive of open-access, to curb the 'race for fish' and thereby migration into—and/or over-investment in—the sector. Methods vary, but typically may include: licence limitation or retirement schemes, buy-back programmes, vessel catch limits, individual effort quotas, and gear and vessel restrictions. These are often combined with the introduction of TACs—Total Allowable Catches. Examples abound. Denmark introduced catch quotas per vessel in the cod, haddock and saithe fisheries in 1989, while the Caribbean Fisheries Management Council prohibits the use of pots/traps, gill/trammel nets and bottom long-lines on specified coral or hard bottom sites across in the region. In the USA, government buy-back programmes costing around US$160 million retired 2,907 permits and led to the withdrawal of 597 fishing vessels across five fisheries over the period 1976-2000, while the Nature Conservancy purchased ten trawling permits in June 2006 as a contribution to the restructuring of the trawl fishery in Morro Bay, California.

Incentive-adjusting devices aim to change the incentive system itself, either by establishing full or partial property rights over the resource. Examples include individual or collective quotas, territorial use rights (TURFS) and co-management schemes, and/or by the introduction of price adjustment mechanisms (such as taxes/royalties/subsidy reduction or removal) to counter resource depletion or overcapacity. Individual transferable quota (ITQs) are increasingly being favoured as a management tool, and have been deployed with varying rates of success across fisheries in Australia, New Zealand, Iceland, Chile and Namibia. Co-management schemes are gaining in popularity too; recent experiences in the Asia Pacific region show that those who exploit the stock are indeed often capable of managing the fishery. Price adjustment mechanisms, such as the removal of the fuel tax subsidy extended to the Bangladeshi fisheries sector, which encouraged boat owners to use more powerful engines, and to fish both longer and deeper in coastal waters, can also play an important role in curbing resource depletion.

Unfortunately, management methods to date have not been an unequivocal success on a number of grounds:

- Inapplicability. While individual quota-based allocative systems function best in single-species industrial fisheries, their suitability for managing small-scale artisanal fisheries and multi-species fisheries (commonplace in the tropics) is highly questionable. Equally, co-management initiatives are dependent on an enabling policy environment, effective institutions, empowered communities, and adequate resources (both fish stocks to manage, and the people and the finances to do the managing)—factors which are not guaranteed to be present in many developed and developing country fisheries.
- Political (Un) Acceptance. The introduction of incentive-blocking (such as vessel catch limits) or incentive-adjusting devices (such as the removal or reduction of subsidies) are likely to be contested by current beneficiaries (among others), and may cause the measure to be rescinded for reasons of political expediency.
- Unintended Policy Outcomes. Buy-out programmes and licence retirement schemes have not always had the impact intended. The removal of active capacity can increase the demand for idle vessels in the fishery and encourage remaining vessel owners to invest further

('capital stuffing') in the belief that the risk of stock collapse has been reduced. A recent review of eleven vessel decommissioning schemes, for example, concluded that in only one—the Japanese Akita fishery—did decommissioning lead to a reduction in total effort expended in the fishery.

While the need to manage fisheries is imperative in the light, it is also clear from the above that current national and international management methods need to be further refined, or re-oriented towards new approaches (such as the ECOST model being pioneered under the aegis of the EU Sixth Framework Programme), if resource sustainability goals are to be met.

FISHERIES AND CLIMATE CHANGE

One further factor which is likely to affect the future evolution of global fisheries is climate change. While the impact of individual climatic fluctuations, most notably *El Niño* events upon pelagic stocks off the Peruvian and Chilean coasts, have been ably documented, the impact of long-term climate variability on fish stocks is less understood.

What is indisputable, however, is that global warming, by raising the temperature of marine and estuarine waters, alters the salinity and acidity of sea-water and the oceanic uptake of carbon dioxide. Moreover, with rising global temperatures contributing to a predicted sea-level rise of between 0.3 and 0.5 metres over the coming century, almost one million square kilometres of coastal land are likely to be flooded, with the concomitant loss of many tidal wetlands, estuaries and mangroves. Fish, as ectotherms, will be severely affected by these changes and we are likely to see a pronounced redistribution of fish stocks and with it, fishing effort, over the coming decades.

SUSTAINABLE FISHERIES

Continued overfishing and environmental destruction endanger more fish stocks, weaken the long-term economic viability of the fishing industry, undermine the stability of coastal communities, and ultimately threaten the contribution of fishing to the global food supply.

The FAO, for example, projects that without a large increase in aquaculture production there will be a potential substantial shortfall of fish and fisheries products by 2010. What should be done? At the most basic level the answer is simple: stop overfishing and protect fish habitat. As Canada's own experience has demonstrated, however, these imperatives can be remarkably difficult to put into practice.

LIMITING EFFORT

Total fishing effort should be limited to sustainable levels for healthy stocks and reduced for depleted stocks to give them the chance to rebuild to sustainable levels. It has been estimated that about 20 million tonnes could be added to the world's annual catch if fish populations were allowed to rebuild.

The World Wildlife Fund (WWF), for example, recommends that effective recovery plans should be a priority of fisheries managers and stresses that target populations and timetables should be driven primarily by the long-term requirements of fish populations and the marine ecosystem rather than by the short-term demands of the fishing industry. Hence, WWF insists that "fisheries management at all levels must be relieved from sweeping political interference aimed at satisfying the short-term needs of the fishing industry." Unfortunately, the WWF goal may not be too realistic as competition for fishery resources, whether between countries, regions, gear sectors, or user groups, often means that allocation of catches is an inherently political process.

Limiting effort to the optimum level, moreover, is much more difficult than it might appear at first glance. In the first place, because of uncertainties in the size of stocks and incomplete knowledge of fish population dynamics, it may not be clear what the limit should be; in the second place, it can be difficult to get an accurate measurement of the true fishing effort. There is also a considerable diversity of opinion on how to limit fishing effort. The WWF recommends that limited-access programmes should form part of comprehensive management plans for each fishery. In fact, limited access is already a feature of fisheries management in virtually all developed countries. In Canada, it was introduced in the early 1980s but did not prevent the collapse of the Atlantic groundfish stocks or prevent B.C.'s salmon fishing fleet from growing far beyond the capacity needed to harvest the catch.

The means of restricting access can often be controversial. Individual transferable quotas (ITQs), for example, are often seen by larger commercial fishing interests as a way to promote good stewardship of the resource and economic rationalization of the industry since holders of quotas have a vested interest in protecting the resource. ITQs are seen by some as an antidote to the "tragedy of the commons" explanation for the overexploitation of fisheries. On the other hand, ITQs are often resisted by small operations and coastal communities; they see ITQs as a way of privatizing the resource that results in a concentration of quotas in the hands of investors who may have no real attachment to the resource or allegiance to coastal communities.

The WWF is critical of measures such as mesh size restrictions and trip limits, describing them as attempts to legislate inefficiency. On the other hand, Safina notes that some regulators have purposefully promoted inefficiency as a way to limit excessive catches and maintain the resource. Examples include laws in the U.S. that require Chesapeake oyster-dredging boats to be powered by sail and the allocation of 52 per cent of the U.S. bluefin tuna quota to the least capable gear: handline and rod and reel. One benefit of the "inefficient" gear is higher employment. In the case of the bluefin tuna, the more labour-intensive gear sector accounts for 80 per cent of direct employment whereas large nets account for only 2 per cent. Sustainability could be improved by reducing bycatches and other incidental damage to juvenile fish and non-target species. In some fisheries, this may mean returning to traditional fishing methods, for example fish traps or the pole and line tuna fishery practised in the 1950s. For other fisheries, it will mean developing and using exclusion devices, such as the Nordmore grate, and more selective gear (such as square, rather than diamond mesh nets).

In principle, it might be possible to enhance productivity by fishing down through the food chain rather than by overexploiting the species at the top. In many coastal areas and coral reef systems, fishing effort has already made this shift so that any gains in productivity would have to come from improved management. In open ocean systems, however, it is not currently economic to move down the food chain.

Scientific research can provide the basis for sustainable fisheries management. Better understanding of species biology, population dynamics, the interactions of different species, and the effects of climatic and other environmental factors will permit more reliable assessments of stocks and more reliable predictions of the way that stocks are affected by harvesting. In Canada, the collapse of the east coast ground fish fisheries has led to suspicion of science. Because science will nevertheless continue to be an essential management tool, however, it will be important to restore a level of trust among scientists, those who fish, fisheries managers and policy makers. Where reliable scientific advice is not available, a more appropriate strategy may be to return to conservative harvesting methods employing low-technology or passive types of fishing gear which are less able to damage stocks.

Economic Viability

The fishing industry currently has twice the capacity needed to harvest the sustainable production of the oceans. This results in estimated annual losses of more than US$50 billion and requires a disproportionately high proportion (46 per cent) of the landed value of the catch as a return on capital investment.

Improving the economic viability of the fishing industry will require a substantial reduction in the capacity of the world's commercial fishing industry in order to match sustainable harvest levels. Such a reduction could restore real profitability. For example, a U.S. study found that by reducing the number of boats by 100, profits from the yellowtail fishery could be increased from zero to $6 million annually.

Compounding the poor economics caused by overcapacity is the fact that many of the vessels over 100 GRT (and presumably also under 100 GRT) are old and inefficient and, according to the FAO, should be scrapped. Because the old vessels are less efficient, they need to catch more fish to break even (let alone make a profit) and the attempt to do this contributes to overfishing. Stimulating new construction, however, would require an improvement in fisheries economics; thus a "catch-22" situation exists.Reducing fishing fleets will not be easy. It has been suggested that one of the most obvious and effective ways to accomplish this would be to eliminate the subsidies that are responsible for much of the overcapacity in fishing fleets. However, this may not go far enough; governments may have to take more active measures to reduce fishing fleets, though these will create an expectation of compensation that will be difficult for governments to meet.

There will also be social and political pressure not to reduce employment and destabilize local economies dependent on the fishing industry. Participation in the fishing industry will, however, have to be reduced to a level that can provide stable and secure incomes within the sustainable limits of the resource. To some extent, the social impacts of capacity reduction may be alleviated by policies that favour labour-intensive over capital-intensive fisheries.

In the longer term, management regimes must be established that eliminate incentives to overcapacity and overcapitalization. Depending on the nature of the fishery and social and economic factors, these regimes may involve market-oriented tools such as individual transferable quotas (ITQs) or the promotion of community-based or other co-management schemes.

Aquaculture

Since 1989, aquaculture production has largely offset the decline in marine capture fisheries and it has the potential to make up an increasingly important percentage of the global food supply, to which it already makes a major contribution. If per capita consumption levels of fish are to be maintained into the year 2010, however, aquaculture production will have to double in the next 15 years. Aquaculture has great potential to increase the production of fish protein, generate economic activity, and provide employment but, as described earlier, it also has the potential to harm capture fisheries and cause social disruption. The growth of aquaculture will therefore have to be carefully managed to ensure that it supplements rather than displaces capture fisheries and that it is based on sound environmental, economic and social principles.

INTERNATIONAL ACTION

Fortunately, there appears to be a growing international consensus supporting conservation of fisheries resources. This chapter briefly reviews some important, recent international agreements.

UN Agreement on Straddling Stocks and Highly Migratory Species

Foreign overfishing on the Grand Banks has helped drive groundfish stocks to the brink of commercial extinction. As a result, a new convention for the protection of straddling stocks and highly migratory species has been a national priority for Canada, which has been working in the UN since the early 1990s to address the problems of high seas fishing.

Agreement was finally reached in New York on 4 December 1995. Signed by 26 member states, the United Nations Agreement on Straddling and Highly Migratory Fish Stocks provides: means whereby members of regional fisheries organizations can take enforcement action against vessels fishing the high seas whose flag states are unable or unwilling to exercise control; compatible conservation measures inside and outside the 200-mile limit; a precautionary approach to fishing; and a compulsory and binding dispute settlement mechanism to resolve disputes concerning high seas fisheries. The Agreement will come into force following ratification or accession by 30 UN member states. As of 20 January 1997, 59 states had signed and nine had ratified or acceded to the Agreement.

FAO Code of Conduct for Responsible Fisheries

The FAO Code of Conduct for Responsible Fisheries, approved in Rome in November 1995, is another important step. The code, described as a "comprehensive moral umbrella," applies to marine and fisheries and deals in depth with fisheries management, fisheries operations, aquaculture development, conservation measures, post-harvest practices, trade, and research. The Canadian government, in partnership with industry, is developing a Canadian Code of Conduct for Responsible Fishing which will take into account the FAO code but will probably go much further.

FAO Compliance Agreement

On 20 May 1994, Canada was the first country to be party to the FAO Agreement to Promote Compliance with International Conservation and Management Measures by Fishing Vessels on the High Seas, which was adopted by the FAO Council in Rome in November 1993. Parties to the FAO Agreement must control fishing on the high seas by vessels flying their flags, in order to ensure that these vessels do not undermine the conservation decisions of international or regional fisheries organizations, even if the Parties are not members of those organizations.

Kyoto Declaration

In December 1995, 95 states met in Kyoto, Japan, to hold the International Conference on the Sustainable Contribution of Fisheries to Food Security. The principles of the Kyoto Declaration, if fully implemented, would bring the world's fisheries much closer to their full potential. These principles include recognition of the importance of fisheries in food security and their social and economic role; steps for the responsible management of fisheries; improvements to food supply through optimum use of harvests and reduction of post-harvest losses; promotion of sustainable and environmentally sound aquaculture; responsible post-harvest use of fish; and ensuring that trade in fish and fishery products does not result in environmental degradation or adversely affect the needs of people for whose health and well-being fish and fishery products are critical.

The world's fisheries have reached, or in many cases even exceeded, the limits of sustainability. At the same time, the world's population continues to increase by approximately 100 million a year and is expected to surpass 7,000 million by the year 2010. The FAO calculates that maintaining current levels of consumption of fish to the year 2010 will require an additional 19 million tonnes of food fish over the 1993 level of 72 million tonnes. It considers this goal feasible if: aquaculture production can be doubled in the next 15 years, and if significant improvements can be achieved in the conservation and management of capture fisheries, through stock rebuilding and more rational harvesting practices, and with the application of food technology to improve utilization of bycatches and small pelagic fish for direct human consumption.

Given all the social, economic and political pressures to keep fishing, together with the environmental effects of fishing and numerous other human activities, this is a daunting challenge and it remains to be seen whether it can be met. The situation also raises the larger question of what happens beyond the year 2010. With the total population of the world projected to rise close to 12,000 million by the end of the twenty-first century, the motivation to keep fishing will be intense.

Fishing is a unique activity. It is the only remaining major world industry that exploits a wild resource for food production; however, it is clear that an unfettered industrial approach to fishing is no longer tenable. Without a fundamental shift in outlook at all levels to one that seriously places the conservation of fish and their habitat as the top priority, there is a serious risk that global fish stocks will continue to decline to a much greater extent than has already happened. The consequences for marine ecosystems, the fishing industry, coastal communities, and, not least, the global food supply could be catastrophic.

IDEA OF A SUSTAINABLE FISHERY

A conventional idea of a sustainable fishery is that it is one that is harvested at a sustainable rate, where the fish population does not decline over time because of fishing practices. Sustainability in fisheries combines theoretical disciplines, such as the population dynamics of fisheries, with practical strategies, such as avoiding overfishing through techniques such as individual fishing quotas, curtailing destructive and illegal fishing practices by lobbying for appropriate law and policy, setting up protected areas, restoring collapsed fisheries, incorporating all externalities involved in harvesting marine ecosystems into fishery economics, educating stakeholders and the wider public, and developing independent certification programmes.

Fisheries have been known to be an important activity throughout the world, and produce more than 100 million tons of fish and fishery products and contribute to human welfare by providing a livelihood to about 200 million people, and protein supply for a billion people. However, with declining stocks as well as several evidences related to fisheries, sustainability issue became very important and has been discussed as the central topic in fishery sciences and industries.

This condition is mainly encouraged by the unfortunate reality that many fisheries are in a state of crisis, and some of them demanding urgent attention. At global levels, fishing industry is a highly adaptive, market-driven and dynamically internationalized sector within the world economy. Its pressure on resources is still increasing, owing to a persistent worldwide upward trend in fish consumption, in concert with human population growth, especially in coastal zones. Global efforts are increasing and have limited the capacity of individual governments to control the pressure over fishing. This problem was associated with a variety of environmental and ecosystem problems including wastage from discards, loss of critical habitats, impact on endangered species, etc..

BIOLOGICAL CRISIS OF FISHERY MANAGEMENT

From this point, fishery sciences itself are progressively switching their attention from single species to ecosystem approach, from micro to macro perspectives, increasing the need for measuring the impact of fishing on natural and man-made systems. Consequently, Cochrane argued that the problems currently experienced in fishery management throughout the world occur in four realms, namely biological, ecological, economic and social crises.

Biological crisis of fishery management started during the last decade after awareness grew about the alarming status of fishery resources. Ludwig for example, drew attention to collapses of fish stocks such as those of the Pacific salmon, the Californian sardine, and the Peruvian anchovy. In 1994, FAO showed, in an analysis, of global fish landings, that there had been reduction in the annual growth in landings in 1980s, and that in 1990 there had been 3 per cent reduction in the global annual catch from the previous year. Garcia and Newton analyzed that this trend continued over the next few years and in between 1990 and 1992, global landings fell by an average of 1.5 per cent per year.

In an economic perspective, fisheries actually exist to meet social and economic demands and one would expect to find that the impact on the resources has resulted in a measurable social and/or economic benefit. Unfortunately, some evidences suggest that the expected benefits have not been in the form of economic gain. Christy estimated that the gross revenues from the total global marine landing in 1989 was US$ 70 billion, giving a deficit balance since the global operating costs were estimated at a level of US$ 92 billion. With an annual capital costs of US$ 32 billion per year, the total deficit in global fisheries was estimated to be US$ 54 billion per year. At a country level, Japan for example, a deficit gross cash flow was also founded for cause of purse seine fisheries. In this type of fisheries, the volume of production was estimated to be lower than its MYS, OSY as well as MEY level. However, due to its high operating cost and low price of fish caught by the purse seine, it resulted in a deficit cash flow. Matsuda called this phenomenon the poverty in large catch. Nevertheless, the recent study undertaken by Tietze shows that in some fisheries, they perform a positive gross cash flow in their global operation. Out of 108 types of vessels from 15 countries throughout the world, 97 per cent had a positive gross cash flow and fully recovered their operating costs. When also considering the annual cost of capital, 85 per cent showed a net profit after deducting the cost of depreciation and interests. However, in terms of sustainability issues, some symptoms of economic crisis in some fisheries should be considered in order to come up with the solution for economic sustainability in the system.

From a social point of view, fisheries are rarely seen as simple tools for generating economic returns. The role of fisheries as a source of employment, particularly in rural or more remote areas, has also widely been given high priority. Moreover, McGoodwin elaborated that fisheries actually are a human phenomenon which essentially are places where human activities are linked with marine ecosystems and renewable resources. Human fishing activities defined the attribute of fishery, since without it there would only be an aquatic realm where various marine species live. Therefore, it has been strongly argued that clearly the fisheries are much more than geographic regions, fishing methods, types of fishing gear, fish species, or economic-domains— it is something more human.

In this regard, social aspects in fisheries would be very important to be understood. One of the major features of fisheries in recent decades, related to the social crisis, has been the introduction of modern fishing technologies and also the increasing globalization of trade affecting the fishing communities. Modernization in fisheries has two faces of which actually one contributed to the welfare of fishing communities, while the other has led to the social problem related to the depletion of fishery resources. According to Crean and Symes, in the North Atlantic and Mediterranean, there has been a decline in the quality of life and standard of living among many fishing communities, as also experienced in some parts of South East Asia. This symptom of social crisis in fisheries should seriously be taken up when we consider the sustainability in this sector.

We argue that the need for measuring and evaluating the sustainability of fishery activities in a system perspective has acquired its highest importance and should be undertaken at various levels involving all aspects in the fishery system. As also strongly argued by Charles that critical concern about sustainability arises not only in terms of catch level or even biomass level, but in all aspects of the fishery: from the ecosystem, to the social and economic structure, to the fishing communities and management institutions, as well as fish stocks themselves. Moreover, the pursuit of sustainable fisheries is best seen as requiring, more than keeping the catch of fish, to keep a level that is not too large. Instead, sustainability can be viewed comprehensively as the maintaining or enhancing of four key components, namely ecological, socioeconomic, community and institutional sustainability indicators.

As discussed briefly in Chuenpagdee and Alder, there have been some methods for analysing and evaluating fishery sustainability, such as RapFish method, FAO Code of Conduct Compliance, and International Instrument Compliance. RapFish is a rapid appraisal technique designed to allow an objective, transparent, multi-disciplinary evaluation. It uses multidimensional scaling, and ordination method to appraise the relative sustainability of fisheries. RapFish relies on a defined-scoring of a number of attributes in five dimensions, namely of ecological, economic, social, technological and ethical. As its characteristics of a rapid appraisal, this method has a slightly top-down approach by employing the defined-score and criteria for the fishery system evaluation. The application of RapFish can be seen, for example, in the Gulf of Maine fisheries, in the North Atlantic Fisheries and recently in the case of Jakarta coastal fisheries, Indonesia.

Similar to RapFish, compliance with the FAO Code of Conduct required multidimensional scaling approach. In this case, four different sets of attributes articulate the clauses in the Code. This evaluation method reflects national management intentions and management practices for each fishery system that is assessed. The application of this method has been carried out in Australian fisheries, Australian multi-species trawl fishery, and Gulf of Maine fisheries. International Instrument Compliance method focuses on measuring either qualitatively or quantitatively the level of compliance with the fisheries management provisions contained within an instrument. In this method, criterion scores ranged from 0 (low compliance) to 3 (high compliance) and the total criteria were limited to approximately six per instrument. An example of this method is presented by Chuenpagdee and Alder in the case of Gulf of Maine fisheries.

It could be said that the above methods rather focus on relatively static approach from the top (top down approach). The fishery in question is just used like the static object which will be evaluated using such methods. They do not involve stakeholders to evaluate their fishery. By these methods, a straight forward analysis could be done but should be treated as preliminary results only. We then argue that it is necessary to confirm dynamic complex systems of the fishery by using a bottom-up approach.

Economic Viability in Fishing Industry

The fishing industry currently has twice the capacity needed to harvest the sustainable production of the oceans. This results in estimated annual losses of more than US$50 billion and requires a disproportionately high proportion (46 per cent) of the landed value of the catch as a return on capital investment. Improving the economic viability of the fishing industry will require a substantial reduction in the capacity of the world's commercial fishing industry in order to match sustainable harvest levels. Such a reduction could restore real profitability. For example, a U.S. study found that by reducing the number of boats by 100, profits from the yellowtail fishery could be increased from zero to $6 million annually.

Compounding the poor economics caused by overcapacity is the fact that many of the vessels over 100 GRT (and presumably also under 100 GRT) are old and inefficient and, according to the FAO, should be scrapped. Because the old vessels are less efficient, they need to catch more fish to break even (let alone make a profit) and the attempt to do this contributes to overfishing. Stimulating new construction, however, would require an improvement in fisheries economics; thus a "catch-22" situation exists.

Reducing fishing fleets will not be easy. It has been suggested that one of the most obvious and effective ways to accomplish this would be to eliminate the subsidies that are responsible for much of the overcapacity in fishing fleets. However, this may not go far enough; governments may have to take more active measures to reduce fishing fleets, though these will create an expectation of compensation that will be difficult for governments to meet.

There will also be social and political pressure not to reduce employment and destabilize local economies dependent on the fishing industry. Participation in the fishing industry will, however, have to be reduced to a level that can provide stable and secure incomes within the sustainable limits of the resource. To some extent, the social impacts of capacity reduction may be alleviated by policies that favour labour-intensive over capital-intensive fisheries.

In the longer term, management regimes must be established that eliminate incentives to overcapacity and overcapitalization. Depending on the nature of the fishery and social and economic factors, these regimes may involve market-oriented tools such as individual transferable quotas (ITQs) or the promotion of community-based or other co-management schemes.

NUTRITIONAL CRISIS AND FISCAL GROWTH IN INDIA

Annual aquaculture growth in India follows the national economy trend. In the same time, 36 per cent of the population lives below poverty line and 48 per cent of the children of 0-5 years old are chronically malnourished. Trade liberalization generated revenues to the national economy from shrimp exports, but at the same time it was responsible for an imbalanced growth and poverty increase at local level. In the agriculture sector, farming of genetically modified organisms (GMOs) by small farmers has compromised their livelihoods, since they cannot afford to pay for GMO grains, inorganic fertilizers, and pesticides. The overall picture implies that nutritional deficiencies and fiscal development can go hand to hand, especially when agricultural products became the new battleground for agro-fuel companies and stock exchange speculators.

SOCIAL CRISIS

The collapses of Bear Sterns and Lehman Brothers opened the Pandora's Box for the globalized economic networks. Whether this is a decaying over-accumulation (capital concentration) crisis, followed by a reproduction crisis of the new high-tech era, or just a creative destruction phase of the previous economic cycle, the current crisis has affected the real economy and society. Total unemployment rate reached 17.5 per cent in the USA and 13.2 per cent of the population (40 million people) live in poverty. This trend is expected to be accompanied by a parallel under-consumption crisis, and negatively affect consumption habits in fishery products.

Currently, the predicted needs for fish food supplies are huge (*i.e.* 270.9×10^6 metric tons for 2050; Wijkström, 2003) and the aquaculture sector is expected to fill the demand gap. However, even if it is possible to overcome the finite character of natural resources, little evidence exists at present that *per capita* supply and consumption of fish products will rise, for example, in the sub-Saharan countries (~less than 8kg *vs* 16.5kg globally), suggesting that food security and accessibility opportunities even to basic goods are not related to the monetary gain-driven nature of the industrial aquaculture and fisheries sectors.

Fisheries in the Twilight Zone

Overexploitation of commercial fish stocks in most of the seas is mainly the result of industrial trawling. The collapse of Northeast Atlantic cod fishery and processing industry in New England during the 1990s was not a sudden event. Early signs of stress in Atlantic cod, halibut and other commercial stocks were evident from 1930s.

The evolution of fishery practices and technology resulted in mass production and decimation of large predators and, subsequently, to extracting fish of the lower trophic levels. It is now widely accepted that overfishing of predators leads to disruption of the energy flows between the food webs, increased by-catch and loss of biodiversity, and, finally, to collapse of the entire vulnerable ecosystems.

The ecological consequences of both 'fishing down the food chain' and 'fishing down the deep' practices will have ultimately serious effects on populations which rely on fisheries. During the 1990s, overfishing and environmental stress in the North Sea caused depletion of commercial stocks and, accordingly, generated increased unemployment, business failures, and rapid structural demographic disturbances in the Faroe Islands. Currently, in developing countries that collectively represent 70 per cent of total catch and 35 per cent of small scale fisher-ies, 22-24 million fishermen and 68-70 million people who work in the postharvest sector are expected to face serious consequences from the fish supply crisis. Political will at the international level currently refects the dominance of fish-ery companies and trade laws over environmental laws, and the future of marine capture fisheries will be ultimately conditioned by political, social and economic factors, including aquaculture.

Metabolic Rifts in Industrial Fisheries and Aquaculture

Sustainable development principles have become the mainstream doctrine. However, the dominant neo-liberal attitude and practice towards the environment (*i.e.* continued economic expansion in order to built capital for investment) has produced irreversible environmental disasters.

Material exchange in biological chemistry and agriculture, where depletion of soil nutrients was observed early (mid 18th century), led Carl Marx to deploy the concept of metabolic rift. Capitalism is the driving force for centuries and organizes social metabolism with nature on the base of capital, production and proft escalation.

However, the equilibrium of these dynamic metabolic processes between humans and nature are frequently ruptured, as constant inputs of fnite resources are required and wastes are accumulated. Such a rupture (metabolic rift) also threatens the stability and biological integrity of aquatic ecosystems. Accordingly, industrial aquaculture ('Blue Revolution') is challenged as a 'quick-fix' to transcend ecological limits and existing metabolic rifts such as

- The development of shrimp farming by the construction of mega-farms both in Central America and Asia. In India, Bangladesh, Thailand and many Southeastern Asia countries, for example, shrimp farms were built on unfit acidic soils after clearing mangrove forests. Moreover, post larvae production is entirely based on wild brood-stock. The use of open farming systems resulted in continuous degrading water quality. Finally, the shrimp sector in most countries has been devastated by the infection by White Spot Syndrome Virus, with serious soioeco-nomic impacts to local communities and economies.
- The unprecedented exploitation of glass eel stocks in European waters was evident during the last 15 years, in order to fulfll the demand by

Chinese farms; it resulted in sky-high prices glass eels (>□1000/kg) and dramatic decrease of natural stocks in European inland waters. Accordingly, eel was registered in the IUCN list of threatened species.

- The exploitation of mature stocks of Mediterranean blue-fin tuna, which have almost crashed and most specimens caught are far below the reproductive age (MacKenzie *et al.*, 2009). Tuna farming for sushi depends solely on state quotas in Europe, as well as on fish caught in the Mediterranean waters of North Africa. Recently, the involved parties (EU, USA, Japan, Canada, etc.) failed to halt tuna fishing during the last CITES meeting, by-passing the urgent alerts of most fishery scientists who suggested zero quotas, while favoring the fishing industry and their lobbies.

TRADE, ENVIRONMENT AND SUSTAINABLE DEVELOPMENT

Globalization has led to increased public awareness of the environmental effects of trade growth and the important developmental implications of issues in the interface between trade and environment. There is a general recognition that increased trade flows that result from globalization have to be accompanied by environmental sustainability and poverty reduction to truly achieve sustainable development. Environmental impact is perceived as an increasingly important factor of production that directly bears on production costs, competitiveness and opportunities in international trade.

If properly implemented, trade liberalization can lead developing countries to access new environmentally sound technologies, goods, services, and production methods. These can facilitate transition to environmentally sustainable production and consumption patterns and augment their international competitiveness. For the first time in the history of the GATT/ WTO, trade and environment issues have become a negotiating subject of global liberalization. Hence the environmental effects of enhanced trade are being much emphasized, and are becoming more and more the centre of public discussion. Issues at the intersection of trade liberalization, environmental protection and economic development have become more closely integrated with globalization.

These are climate change and biodiversity; new environmental, health, and food-safety requirements; and access to environmental goods, services and technologies, and related sustainable production methods. They will pose formidable challenges for the international community in the years to come as any attempt to reduce poverty will have to take the natural environment into consideration.It is the poor who are the most dependent upon the natural environment to meet their daily food, health, livelihood, and shelter needs. Thus the effects of environmental degradation are felt most immediately and keenly by the poor.

TRADE, CLIMATE CHANGE AND SUSTAINABLE DEVELOPMENT

The international community has now reached a consensus regarding the fact that the increasing emissions of greenhouse gases such as carbon dioxide and methane – most of which are linked to the human use of fossil fuels – are causing changes in global climate systems. Climate change currently poses one of the greatest risks to environmental, social and economic development globally. Private and public responses to the climate crisis are bringing significant changes in several economic sectors, especially related to energy. Some of the recent trends in the energy sector are outlined in Chapter VII.

This section highlights the broad range of relationships among climate change, trade and development: how trade policy might impact on climate change through economic transformation; how significant competitiveness and market access concerns may be affected; how climate change may physically impact economic structures, in particular in agriculture and services, as well as infrastructure; and how trade rules interact with measures for climate change mitigation.

The UN Development Programme's Human Development Report 2007/2008 focuses on potentially dramatic impacts of climate change upon agricultural production and food security, water stress and insecurity, rising sea levels and flooding, ecosystems and biodiversity, and human health.

The Intergovernmental Panel on Climate Change outlines mitigation options for the following sectors: energy supply, transport, buildings, industry, and agriculture, forestry, and waste - reflecting the extent to which climate measures are affecting nearly every aspect of the economy.The new sense of urgency behind efforts to curb global warming may provide the stimulus for a more proactive approach to integrating trade policy within sustainable development strategies. Global concerns on the impact of climate change have emerged as a key development theme with globalization.

Impacts of Trade and Investment upon Climate Change

International trade may impact climate change in a multi-faceted way through its scale effects, resulting in increased economic activity; composition effect leading to changes in the structure or patterns of economic activity; boost and changes in technology; and direct GHG emission effects, *inter alia*, from increased maritime, truck and air transport. The scale effect of trade will generally have a negative climate change impact, because higher production of most goods and services will generate more GHG emissions. The composition effect of trade liberalization tends to shift production to goods and services in which countries have a comparative or absolute competitive advantage. Depending on national policies, this might lead to a more or less carbon-efficient economy.

The overall outcome, however, could be a global reduction of GHG emissions, provided there is internalization of the environmental costs of GHG emissions. The technological effect of enhanced trade and investment flows can make a significant contribution to material and energy efficiency, and thus to climate change mitigation. Conversely, more trade and investment generally lead to directly higher GHG emissions from increased maritime, truck and air transport, in addition to higher electricity consumption by global computer and telecommunications networks.The right mix of specific trade and investment policy measures can optimize multi-faceted impact of trade and investment liberalization on climate change.

In this regard, tariff liberalization for renewable and clean conventional energy and related equipment, as well as energy-efficient goods or inputs for energy- and carbon-efficient production processes is one promising cluster of trade policy tools. Another is the reduction or removal of subsidies for conventional energy sources and energyintensive sectors.

A third cluster is the use of technical requirements and standards to encourage carbon-efficient modes of production and consumption. Fourthly, government procurement can be used to encourage consumption and investment in low-carbon goods and technologies. Last but not least, investment policies can be geared to gradually redirect investment into carbon-efficient sectors and simultaneously enhance carbon efficiency in energy-intensive industries.

This implies greater opportunities for energy-efficient and carbon-neutral industries and stimulating technological innovation. Also, changes in the energy mix will often support local energy and development needs more effectively than fossil fuel imports. The Clean Development Mechanism of the Kyoto Protocol offers significant opportunities for attracting and directing investment into carbon-efficient or carbonneutral areas, including changes in the energy mix.

In order to restructure markets towards carbon neutrality, consumers, corporations, and governments need to consider certain increased costs as medium to long-term investments. Businesses directly involved in GHG-intensive activity need to be supported to make the necessary transition to minimize dislocation in the interest of broader societal benefits.

Developing countries that invest private capital and direct their public policies towards climate-friendly products will stimulate domestic innovation, reap the advantages of technological "leapfrogging", and are likely to increase their export potential. Developing private and public sector strategies with climate change in mind also often offers local environmental protection and other benefits as well. Countries that fail to do so risk consolidation in dirty industries on the lower end of the value chain, and damage the environment.

Possible tensions between trade law and attempts to address climate change

Although the UN Framework Convention on Climate Change and the Kyoto Protocol have no specific trade obligations, they have significant trade implications as they aim to modify the carbon impacts of the ways in which goods and services are produced and consumed. The interface between trade rules and climate change concerns the WTO disciplines on tariffs, technical barriers to trade, government procurement, subsidies, investment policies, and border tax adjustment.

The growing importance of national measures to address climate change, the UNFCCC as an element of international economic governance, and climate change as a factor in international commerce will heighten the importance of the interface between trade and environment policy but does not pose an inherent conflict. Rather, it does necessitate enhanced coordination among policymakers. The principle of differentiated level of obligations among the parties according to their different stages of development – adopted in Agenda 21 in 1992 – is the basis of the UNFCC and the Kyoto Protocol. There are prospects for evolving post-Kyoto commitments frameworks with a potential for further engagement by some developing countries. The United Nations Conference on Climate Change in Bali resulted in the adoption of the Bali roadmap. The Bali roadmap charts the course for a new negotiating process to be concluded by 2009. This will ultimately lead to a post-2012 international agreement on climate change. There is also increasing interest in ensuring access to low-carbon technologies and to additional financial resources for implementing climate change policy.

Competitiveness and market access concerns

The competitiveness concern arises regarding two issues. On the one hand, the likelihood that a country that takes stronger climate change mitigation measures might put its companies or industries at a disadvantage relative to foreign competitors in countries that do not adopt similarly strong measures. This may lead to a "carbon leakage" problem, where strong mitigation measures may encourage companies to relocate to other countries. On the other hand, there is concern that even among countries taking similarly strong climate change mitigation measures, there is unfair competition due to differentiated employment of such measures. The World Bank finds that "there is some evidence – although it is not very pronounced – of leakage of carbon/energy-intensive industries to developing economies that could be attributed to more stringent climate change policies and energy efficiency standards."Sectoral characteristics also matter, *i.e.*, how energy intensive is the economy and to what extent companies and the sector are in a position to pass on cost increases to customers.

Developing countries face a significant challenge due to their industrial structure and its carbon intensity. Investment in energy-intensive industries in developing countries may expand, resulting from domestic needs for industrialization, energy security, physical-infrastructural development, but also redeployment of carbon-intensive industries from developed to developing countries. For example, "environmentally-sensitive industries" represent a growing share of exports for several Latin American countries.Within these sectors, developing countries often concentrate on the bulk market segment where higher carbon/energy prices are difficult to pass onto consumers. A potentially significant problem for developing countries is the competitiveness and market access impact of technical measures to trade, caused by requirements on energy efficiency product and process standards or related eco-labelling programmes. In its recent study, based on a simple two- country trade model, the World Bank singles out energy-efficiency standards as likely to have the most significant trade impact of all trade policy tools for climate-change mitigation.

Impact of climate change on trade and investment

Climate change will have significant implications on trade flows arising from its impact on agriculture, forestry, trade-related physical infrastructure, and services such as tourism. Weather extremes and related natural disasters can disrupt specific sectors, notably agriculture, and negatively impact infrastructure, in particular along coast lines. Tourism is likely to be impacted by weather conditions and fundamental ecological changes. Rising temperatures are also likely to modify the competitive advantage in agriculture based on ecological factors.

The Intergovernmental Panel on Climate Change forecasts imply gradual but colossal shifts in production patterns, cultivated crops and yields, the spread of pests and diseases, as well as accelerated desertification and droughts. The introduction of climate response measures through the emerging carbon market and the Kyoto Protocol will have trade and development implications as they are introduced in several sectors of the economy, such as transportation, energy use, electricity generation, agriculture and forestry. Many developing countries, in particular the small and vulnerable ones, will be particularly hard hit by these climate change impacts. For example, sea-level rise is causing enhanced soil erosion, loss of productive land, increased risk of storm surges, reduced resilience of coastal ecosystems, and raising attendant costs of responding to and adapting to these shocks. Countries in temperate zones are likely to be far less affected or may even benefits from a longer vegetation period, more cultivable crops and higher yields. As a result, international trade patters in agriculture will change over time because of different supply and demand patterns, as well as yields.

BIODIVERSITY, TRADITIONAL KNOWLEDGE AND TRADE

The international approach to the protection of biodiversity continues to focus on innovative ways to promote sustainable use, bringing economic, social and environmental benefits to nations and their people. Biodiversity-rich countries are taking advantage of new trade and investment opportunities for biodiversity products and services in the emerging market, with increasing participation of the local private sector. UNCTAD's BioTrade Initiative has estimated that the world market for natural ingredients used in the cosmetic and pharmaceutical sectors amounts to over US$1 billion annually.

Greater scientific certainty, public awareness, growing trade and investment activity levels, as well as the availability of statistical data on environmental and economic losses associated with certain patterns of development, have all led to more discussions and the promotion of more pragmatic policy options aimed at sustainable development. Harnessing the knowledge-for-development focus will require assisting developing countries to benefit from their own resources: their rich traditional knowledge, innovations and practices.

TK is the main asset of the poor who use it to derive goods and services from their natural environment. Yet TK is being lost at alarming rates worldwide, as globalization and environmental degradation are accelerating the break-up of traditional communities and endangering livelihoods. There are also concerns that TK is being inappropriately exploited and patented by third parties without the consent of the original holders of that TK, and without the fair sharing of resulting benefits. There is need for concerted action at the national, regional, and international levels to redress this. A key example is organic agriculture, whose production systems are built on a synthesis of local TK and the results of modern research. In this way, producers who use local varieties adapted to local conditions can achieve higher incomes and greater security than would be the case with conventional agriculture. Further, the protection of TK as intellectual property is an important means of harnessing the potential benefits of TK for trade and development gains by the owners themselves.

NEW ENVIRONMENTAL, HEALTH AND FOOD SAFETY REQUIREMENTS AND MARKET ACCESS

- An important trend at the trade-environment-development interface is the growing impact of new environmental, health and food-safety requirements on the access of developing country products to key export markets. The proliferation of private voluntary standards on EHFSRs and sustainablility standards in international trade, and their impact on market access and national development of developing countries is a major concern. In contrast to the proliferating standards, there is a dearth of empirical knowledge both about their impact as well as about successful adjustment strategies to these standards taking into account national developmental priorities. Four developments are particularly challenging:New EHFSRs are becoming more stringent, frequent, complex and multidimensional. This constitutes both serious challenges as well as opportunities for export competitiveness, sustainable production, and consumption methods at the national level.

- There is a growing trend towards the 'privatization' of many EHFSRs, with voluntary requirements imposed by the private sector co-existing and inter-acting with mandatory governmental requirements. Governments set product characteristics, product-related processes, and production methods; the private sector and NGOs follow by imposing specific non-product-related PPMs to meet the product characteristics. As it is open to question whether or not private standards fall under the WTO disciplines, they pose serious challenges in terms of justifiability, transparency, discrimination and equivalence.
- Besides their function of providing technical quality-assurance, private standards often play a governance role in global supply chains, leading to significant dependencies and the shifting of costs and risks away from buyers, often to the disadvantage of producers/exporters in developing countries.

The new bread of private voluntary standards, but also some sustainability standards of NGOs, poses particular challenges for small farmers in developing countries. It is not so much the lack of quality or productivity of small growers, but the enhanced management and coordination costs in implementing and complying with PVS that are causing very high recurrent adjustment costs, on average about 20 per cent of turnover or up to 50 per cent of total income of small farmers, often causing massive drop-outs of small producers from PVS compliance schemes. According to a recent study on horticultural exports from Ghana, Kenya, United Republic of Tanzania, Uganda and Zambia to the United Kingdom, over 50 per cent of small producers participating in PVS compliance schemes have dropped out of these schemes between March 2005 and September 2006.

The mushrooming of new EHFSRs has to be dealt with in a proactive and coordinated way. This requires the development of national adjustment strategies that minimize the costs and maximize the benefits of the new requirements.

It also requires actions to seize production and export opportunities including through 'front-of-pipe' solutions on production processes, materials for environmentally friendly goods and services. These would include organic agriculture products, other biodegradable products, natural colorants and flavours.

Of particular importance is conceptual clarity on the design of national programmes on Good Agricultural Practices that in a modular way allowing producers to meet national and regional requirements with buyer recognition in lucrative overseas export markets. In these national GAP programmes, governments need to pay special attention to support the participation of small producers through supportive/flanking measures, which lower adjustment costs, provide bridging funding, support the creation or continuation of stable and wellmanaged groups of small producers, and provide improved extension and training services.

DEVELOPING ENVIRONMENTALLY PREFERABLE PRODUCTS, SERVICES AND PRODUCTION METHOD

International public attention on the problems caused by climate change, material and pollution intensity of economic growth and unsustainable life styles as well as the pressure from new EHFSRs have heightened the interest in environmentally preferable products, services and production methods. These are the strategic markets of the future.

Developing countries need to identify market niches and the opportunities open to them as well as the policy initiatives needing to be launched in time to turn these opportunities into reality. The growing consumer demand for environmentally preferable products presents new opportunities for those producers and countries that can produce them in more energy-efficient and environmentally friendly ways, especially if they can effectively communicate this to consumers.

A prime example of this is the rapid expansion of organic agriculture markets: global growth rates have been over 12 per cent over the past decades, and compare favorably with the overall agriculture market growth of only 2-4 per cent. In addition to the economic advantages accruing from premium prices and expanding sales, organic agriculture offers developing country producers an array of environmental, health, social, cultural, and food security benefits. Other examples of EPPs include energy-efficient electronic goods and certified wood products. Yet even here, differing standards can become obstacles to trade. Thus, further harmonization and equivalency are needed to fully reap the gains in trade and sustainable development.

INTERNATIONAL TRADE AGREEMENTS FOR MANAGING THE TRADE AND ENVIRONMENT INTERFACE

In the Doha Round, and for the first time in the WTO history, trade and environment has become a negotiating subject. The main challenge to the negotiations provided for in paragraph 31 of the Doha Ministerial Declaration on environmental goods and services is to make three main objectives – environmental sustainability, development and trade liberalization – converge in a mutually supportive way. The current positions span a wide range of approaches. On the supply side, there is the very pragmatic approach of putting forward self-defined lists of environmental goods. On the demand side, there is the environmental project approach, which seeks to strengthen the hands of the individual countries reflecting their divergent environmental situations and developmental priorities. Still other approaches seek to bridge both sides in the negotiations. Irrespective of which negotiating approach prevails, it will have far reaching implications in the longer term.

The risk lies in the absence of criteria, which may lead to a precedent, an inadequate introduction of this subject matter in trade liberalization rounds, with consequences for the following rounds and tendency to deal with the issue on the basis of negotiating power. Climate change policy highlights the growing interface between energy and environmental goods and services. Goods, equipment and technologies used in conjunction with renewable energy sources - renewables - are one case in point. The strength of the international commitment on climate change may play a catalytic role and influence the modalities for cooperation, including in the WTO, be it as a separate "WTO climate initiative" or in the context of the negotiations conducted under the mandate provided for in paragraph 31 of the Doha Ministerial Declaration regarding the reduction or, as appropriate, elimination of tariff and non-tariff barriers to environmental goods and services..

As climate change does not form part of the negotiating history in the Doha Round, there is a need to carefully manage the interplay between the on-going work and any new initiatives. There is a growing consensus, however, that the negotiations on environmental goods and services should include renewables, and possibly technologies for cleaner utilization of conventional energy sources such as natural gas-driven turbines, low-emission coal combustion, and carbon capture and storage.

They could also include climate positive goods such as biofuels and energy-efficient construction materials and appliances or even goods derived from more GHG-efficient processes and production methods. The expansion of product coverage in the negotiations to include these goods is not without problems. While the existing HS can capture most renewables, the ubiquitous nature of some goods and their component parts means that the dual use problem will remain over and above what could be sorted out by a greater specificity in the tariff codes. As for GHG-efficient goods or goods produced in a GHG-efficient way, there are simply no HS codes to match, not to mention that climate or energy efficiency is a moving target. The inclusion of goods derived from GHG-efficient processes and production methods is especially problematic as it may dramatically increase the scope for protectionist measures.

The relative importance of tariffs and non-tariff barriers is another sticky point. Lowering the tariff reductions may well be a simpler task, but NTBs are considerably more important for the liberalization to be commercially meaningful. Another important issue is that climate positive technologies and the export of related goods tend to be concentrated in developed countries.

It is important to balance market opportunities brought about by climate positive technologies and access to these technologies by developing countries. "Technology will play an essential role in our collective response to climate change".The ultimate goal is to capture the "public goods" nature of innovation and international trade. More than 50 countries with 80 per cent of the world's population now have measures in place, both mandatory and voluntary, for energy efficiency. And yet there is a significant gap between expectations and actual impacts. At the same time, energy efficiency standards affect trade flows and market access. As the marketplace for energy and environmental goods and services becomes increasingly global, so too is the need to ensure cooperation in the development of these standards. Financial flows and official development assistance targeting climate positive technologies may play a catalytic role in the development of renewables in developing countries through increased trade, investment and technology transfer. Some developed countries are involved in a number of projects in developing countries. The international consensus on climate change provides an important incentive framework for various types of cooperation arrangements. Countries increasingly use bilateral and regional trade agreements to manage the trade and environment interface.

A variety of instruments has been deployed, ranging from environmental chapters and side agreements to consultation, cooperation, and exception clauses. Among the OECD members, Canada, EU, New Zealand and the USA have included the most comprehensive environmental provisions in recent RTAs.

The 'Global Europe' communication of November 2006 announced a set of new bilateral negotiations by the EU. Environmental concerns have a very important role in these negotiations, which will be seeking substantial commitments from both sides, with possible market access and development assistance incentives. Among non-OECD countries, Chile's efforts to include environmental provisions in its trade agreements are particularly noteworthy. Few trade agreements among developing countries include environmental provisions like ASEAN and MERCOSUR.

FOSTERING AN ENABLING ENVIRONMENT: PROMOTING TRADE AND ENVIRONMENTAL SUSTAINABILITY

The interface of trade-environment-development with globalizations necessitates a transition to environmentally sustainable production and consumption patterns as well as international competitiveness. Trade policies and trade liberalization should facilitate access to new environmentally sound technologies, goods, services, and production methods. In meeting this challenge, dealing with climate change has major trade and development implications that the world as whole has to address. Three sets of policies are accordingly being developed at the international level to ensure a coherent approach that will minimize the detrimental effects and maximize possible opportunities.

One set of policies concern 'Cap-and-Trade' policies on carbon pricing, taxation, emissions trading, and regulation. These policies will ensure that people face the full socio-economic costs of their actions. A second set of policies promotes home-grown technological solutions and incentives that will drive the development and deployment of a wide range of low-carbon and high-efficiency products and services.

A third set of policy measures will aim to remove barriers to energy efficiency, and to inform, educate and persuade individuals on what they can do to respond to climate change in each sector of the economy. Effective action to counter climate change impact requires a global policy response. These typically involves: mitigation, or the reduction of greenhouse gas emissions; and adaptation, or by 'climate proofing' economies. If the future global climate regime is going to contain emission reduction commitments involving some developing countries, developed countries should assist them in capacity-building, technology transfer, and adaptation. In parallel, the impact of efforts made towards reducing emissions in those developing countries which are highly dependent on the production/export of fossil fuels should also be examined.

Special support should be provided to them. The long-term nature of the climate change problem makes technological change a central issue in policy considerations. Specific financing mechanisms should be made available to them to help in the process of developing and adopting new energy technologies. Renewable energy sources comprise an important means to reducing climate change.

Developing countries have two advantages contributing to the competitiveness of their renewable energy sources. They tend to have strong renewable energy resources and a lower costs profile for the production of equipment, components, and biofuels. These two factors point to the scope for trade and cooperation in renewable energy. However there are major considerations that still have to be addressed. These include tariff barriers affecting trade in renewable energy, market deployment policies for renewable energy and other carbon-reducing energy technologies. Financial flows and official development assistance targeting climate positive technologies may play a catalytic role in the development of renewable energy sources in developing countries through increased trade, investment and technology transfer. Some developed countries are involved in a number of projects in developing countries.

The Kyoto Protocol provides an important incentive framework for various types of cooperation arrangements. Addressing new EHFSRs is another emergent issue on the international agenda in the area of trade and environment. The concern is how to design appropriate proactive adjustment strategies to address new EHFSRs. This requires conceptual clarity, a good understanding of the role of supportive or flanking policies, effective public-private partnerships as well as policy coherence at the national level. At the initiative of several developing country members, the WTO SPS and CTE Committees have recently begun discussing the salient issues of private sector standards and WTO disciplines.

The development of regional standards by developing countries – as part of proactive adjustment approaches – is a worthwhile initiative because, apart from facilitating access to overseas markets, such standards can also ease regional trade. Liberalization efforts in the WTO on environmental goods and services should be considered in conjunction with the possibilities for supporting and financing these efforts, and to make them commercially, financially and technically viable. So far, no institutional linkages have been established between the negotiations and the different fora dealing with development finance and assistance.

There is a need to promote coherence in the negotiations between the WTO and other environmental infrastructure projects financed by multilateral financial institutions, especially in terms of meeting financial needs and building capacity. The key driver in introducing environmental provisions in RTAs is ensuring a level playingfield among the parties. This can be done by giving a legal expression to a commitment to maintain high levels of environmental protection.

Another motivation is to enhance cooperation in environmental matters. The fact that an increasing number of RTAs serve as a framework for cooperation in environmental matters does not necessarily make such cooperation traderelated. In fact, most cooperation agreements are not trade-related. They have typically taken place among countries with shared ecosystems. Indeed, a lot of regional arrangements for environmental cooperation pre-date the respective RTAs. Finally, there seems to be a missing link between international trade negotiations and the need to be responsive to broader development goals, such as the MDGs. There is a need to look at the technical issues arising in the negotiations on trade and environment from a broader development perspective. The MDGs are one case in point.

There are many potential target areas to be derived from MDGs such as the supply of drinking water, drainage systems, sanitation, the disposal of sewage, waste disposal, and the development of renewable energy sources. The choices the WTO Members make could also be linked to multilateral environmental agreements. Bringing a development dimension to negotiations on environmental goods and services is important to promote sustainable development.

FISHING BUSINESS FOR SUSTAINABLE DEVELOPMENT

The Millennium World Food Summit, 2000 held in New York set an ambitious target of halving the percentage of hungry people by 2015. The WFS overall objective was "... to eradicate hunger in all countries, with an immediate view to reducing the number of undernourished people to half, no later than 2015". This represented a target goal of 412 million undernourished people; down from 824 million (estimated) in 1996 it self. During the beginning of the new millennium, *i.e.*, from 2000 to 2002 it was estimated that 852 million people were undernourished world wide. The global population in the mean time has increased in several folds.

It is relevant to mention that food security is a complex phenomenon that relates more to economic development and poverty than to increasing production *per se* (FAO, 2005b).The produced food must therefore be available to the people at a reasonable price and legislation must prohibit illegal storage and black marketing, etc. So the concern of providing food and nourishment to the exponentially growing global populace is a challenge that should be tackled by the policy makers with a sustainable approach.

Intensive aquaculture practices in this context is seen as an alternative to meet the widening gap in global rising demand and decreasing supply of nutritive food products. It has vast potential in providing livelihood security as well as fulfilling the nutritional requirements of the growing population and becoming an increasingly important food production process. The strata of Indian problem, *i.e.*, from providing food and nutrition to one of the largest population of the world, generation of employment for the youth and socioeconomic development of the villages can be answered partially to an effective extant by means of integrated management practices of natural resources such as intensive aquaculture.

WHY AQUACULTURE

About one billion human beings worldwide depend on fish as their primary means of animal protein. Fish contributes around 50 per cent of total animal protein consumed in Indonesia, Japan, India, Sri Lanka, Bangladesh and Cambodia. Aquaculture and fisheries can meet the huge demand for nutritive, value for money food and related products.

Consumption of both high and low value food fish is growing in the developing world and the developed countries as well (FAO, 2005). In 2004, per capita food fish supply was estimated at 13.5 kg. On an average fish provided more than 2.6 billion people with at least 20 per cent of their average per capita animal protein intake. Global consumption of fish per capita in 2005 was at a peak of 16.6 kg per capita. The share of fish protein in total world animal protein supplies grew from 14.9 per cent in 1992 to a peak of 16.0 per cent in 1996. Notwithstanding the relatively low fish consumption by weight in low-income food-deficit countries (LIFDCs) the contribution of fish to total animal protein intake was significant at about 20 per cent and is probably higher than indicated by official statistics in view of the unrecorded contribution of subsistence fisheries (FAO, 2007).

Growing domestic demand within developing nations for fish is driven by good governance n the sector ensuring easy availability and economical pricing of per Kg fish protein. In the rich world fish and related products are increasingly seen as a healthy luxury food due to regular media reports of newer scientific findings about nutritional benefits of fish and related products. Fish products are now actively marketed by mega life style stores in the biggest business districts of the world. In contrast Asia and many other developing countries still serve fish as an important part of the staple diet.

The fish and related food products are generated either by capture fisheries or by culture fisheries (aquaculture).Culture fisheries is more sustainable than of capture fisheries in ecological viewpoint. Aquaculture market is set to witness the growth of about 4 per cent or more in recent times. Capture fisheries and aquaculture supplied the world with about 106 million tonnes of food fish in 2004, providing an apparent per capita supply of 16.6 kg (live weight equivalent or LWE). Of this total, aquaculture accounted for 43 per cent. In 2005 aquaculture production was at a record 47.5 million tonnes, or 34 per cent of total fish production.

If calculated on the basis of fish for human consumption only, aquaculture production constitutes 44.6 per cent of the total. Global aquaculture and fisheries market crossed 67 million tons approximately in 2008. According to reliable media reports Global aquaculture and fisheries market is expected to exceed 123 million tonnes by the end of 2009. An FAO study projects that capture fisheries could produce some 12 million tonnes more by 2015, compared with 2005 levels, and that aquaculture production could reach 66.8 million tonnes by then.

Over the years the magnitude at which fish and related food products are preserved and processed, has also increased in several folds through out the world. This nevertheless is a clear indication of serious business, as the global community is engaging actively in the food fish industry. In 2004, about 75 per cent (105.6 million tonne) of estimated world fish production was used for direct human consumption, and the remaining 25 per cent (34.8 million tonne) was processed into feeds, mostly fishmeal and oil (FAO, 2007), besides 7.3 million tonne discarded. Some 61 per cent (86 million tonne) of the world's fish production (2004 figures) underwent some form of processing, and 59 per cent

(51 million tonne) of this processed fish was used for manufacturing products for direct human consumption in frozen, cured and canned form.

The rest went for non-food uses. Unlike many other food products, processing fish does not necessarily increase the price of the final product, and fresh fish is often the most highly priced product form. Freezing is the main method of processing fish for food use, accounting for 53 per cent of total processed fish for human consumption in 2004, followed by canning (24 per cent) and curing (23 per cent).

In developed countries, the proportion of fish that is frozen has been constantly increasing, and in 2004 accounted for 40 per cent of total production. In comparison, the share of frozen products was 13 per cent of total production in developing countries. Utilisation of fish production shows marked continental, regional and national differences. The proportion of cured fish is higher in Africa (17 per cent in 2004) and Asia (11 per cent) compared with other continents. In Europe and North America, more than two-thirds of fish used for human consumption was preserved in frozen and canned forms. All these statistics had an upside move during the next couple of years.

AQUACULTURE AND ITS ECONOMICS

During the year 2003 the export value of world trade in fish was US$63 billion which was more than the combined value of net exports of rice, coffee, sugar and tea. Half of global fish trade comes from developing countries, while global consumption increased by 21 per cent between 1992 and 2002. In 2004, total world trade in fish and fishery products reached a record value of US$71.5 billion (export value), representing 23 per cent growth relative to 2000 and 51 per cent increase since 1994. Preliminary estimates for 2005 indicate a further increase in the value of fishery exports. In real terms (adjusted for inflation), exports of fish and fishery products increased by 17.3 per cent during the period 2000 to 2004, 18.2 per cent during 1994 to 2004 and 143.9 per cent between 1984 and 2004. Fish is traded widely, so today it can be said that fish from all corners of the world can be found on the international market.

In 2004, about 38 per cent of all fish produced (LWE) was exported as various food and feed products. Developed countries exported some 23 million tonne of fish (LWE) in 2004. Although a part of this trade may be re-exports, this amount corresponds to about 75 per cent of their production. Exports from developing countries (30 million tonne LWE) totaled around one-quarter of their combined production, but, remarkably, the share of developing countries in total fishery exports was 48 per cent by value and 57 per cent by quantity (FAO, 2007).

For the developing world, fish exports have become an ever more important source of foreign exchange. Fish is classified in the world trading system with industrial products, and thus carries very low tariffs compared with agricultural goods. Some 38 per cent (by volume) of all fishery production enters international trade, with over half of that originating in developing countries. The globalisation of fisheries and the wide participation by both developed and developing countries in world fish trade is encouraging for the aquaculture communities. (Valdimarsson, 2007) and the explosive growth in this sector over past decades has been accompanied by a boom in international fish trade (FAO. 2005d) (Valdimarsson, *et al.*, 2005).

GENERATION OF EMPLOYMENT AND LIVELIHOOD

Aquaculture and Fisheries sector over the years has also contributed significantly in employment generation. According to 2007-08 estimates of the World Bank, the livelihood of about 200 million people relies on fisheries, aquaculture and associated activities. Small scale and large scale fish farms were actively involving skilled and non- skilled workforce. The numbers of small scale or subsistence fisheries has been constantly rising over the last decades through out the world and were estimated to be over 41 million in 2004, including some 11 million fish farmers. Many a times the same individuals are engaged in both (FAO, 2007). The contribution of the small-scale fleet to fish for human consumption may be as high as 50 per cent. However, in industrialised countries the number of individuals engaged in organised or unorganised fishing business was much lower in comparison and estimated to be about 1 million.

ECOLOGICAL SUSTAINABILITY OF AQUACULTURE OVER CAPTURE FISHERIES

The huge market for fish and related products also has environmental concerns in addition to economical and social challenges. The plethora of the demand is causing strains regarding trade policies as well as conservation of biodiversity which has exacerbated the need for fisheries management capable of scientific and social innovations and to keep fish catches within sustainable limits. In recent times the global community has emphasized much on sustainable and inclusive development to bring about fundamental changes and growth in all spheres of human life. Although the overall growth in the fisheries sector is very much encouraging, but sustainable aquaculture practices and innovations are therefore a current need of the day. Coastal and inland capture fisheries are putting pressure on fish stocks worldwide (Charles, A.T 2001).

Maximum Sustainable Yield, an important concept in aquaculture has been developed to avoid overexploitation of fishing sites. The exploitation pressure if crosses certain threshold will lead to an irreversible damage on the ecosystem. Ecosystems as such cannot be managed, but only the human activities of exploiting them (FAO, 2003a) can be regulated. Over the last two decades, the marine fishery resources of the world have been increasingly subjected to overexploitation (Bodiguel et al, 2009) mainly due to detrimental fishing practices leading to environmental degradation (FAO.2005d). This phenomenon has adversely affected a majority of fisheries worldwide, with very severe consequences in terms of resource destruction due to loss of breeding grounds, massive economic wastage due to sudden fall in the fish catch and increasing social cost as well as food insecurity.

Currently, FAO estimates that more than 25 per cent of the all fish stocks on which it has information are over fished, depleted or recovering from depletion, whereas 52 per cent of the stocks are fully fished and only 23 per cent of the stocks could produce more. The worldwide wild capture fisheries potential has reached its limit and so the huge demand for aquatic protein will be satisfied by aquaculture. Therefore the farmers and fishers should be encouraged to engage in responsible fishing following good fisheries governance (Sinclair *et al.*, 2002). Intensive aquaculture as well as innovative protocols should be developed with a collaboration of all the governments with an integrative approach by combining scientific and technological advancements so that the global biodiversity resources are not exhausted or over exploited.

AQUACULTURE AND FISHERIES IN ASIATIC REGION

Asia, the epicenter of the global aquaculture industry, accounts for over 90 per cent of the global aquaculture production quantity and about 80 per cent of the value. Asia-Pacific region forms the major fisheries and aquaculture market in terms of production. Among the top ten countries where fish plays an important role of animal protein supply, four are African countries and five are from the Asiatic region. During 2006 the Asia-Pacific region lead production of Aquaculture and Fisheries with an estimated share of about 60 per cent of the global out put, as stated by Global Industry Analysts, Inc a global economics research firm. Asia-Pacific is forecast to maintain leadership position with a CAGR of over 3.3 per cent during 2006-09. Europe and Japan represent the other major regions for Aquaculture and Fisheries. Europe is projected to account for over 19 million tons of aquaculture and fisheries product by the year 2010.

AQUACULTURE AND FISHERIES IN INDIA: THE CURRENT SCENARIO

The National Agriculture Policy 2000 of Government of India accorded high priority to increase protein availability in the food basket and generation of exportable surpluses. India being a developing nation, is constantly exploring ideas for an inclusive sustainable development for the people. Fisheries sector plays a very important role in the socio economic development of India. Indian fisheries has made great strides during last five decades with the production levels increasing from 750,000 tonnes of fish in 1950-51 to 6.4 million tons in

2005-2006, of which the contribution from the inland sector is around 3.3 million tons (51.6 per cent of the total) compared to 3.10 million tons (48.4 per cent) from the marine sector. India's contribution to global fish production increased from 3.26 per cent in 1985 (Alagarswami, K. 1995) to 4.41 per cent in 1997 itself.

The contribution of fisheries to the gross domestic product (GDP) and agriculture GDP has been estimated to be 1.2 and 4.2 per cent, respectively. India ranks second in world inland fish production, next to China. The growth rate of inland and marine sector at present is 6.6 and 2.2 per cent, respectively. During 2006-07 total fish production of 6.57 million metric tonnes had nearly 55 per cent contribution from the inland sector and nearly the same from culture fisheries.

The growth rate during the tenth five-year plan (2002-2007) for inland and marine fisheries were 8.0 per cent and 2.5 per cent, respectively. According to Foreign Trade Statistics of India (Principal Commodities and Countries), DGCI & S during the recent years there has been steady growth in the revenue earned by India from marine products. During 2000-01 total revenue earned from marine products was 6296.00 cores INR whereas during 2006-07 it was 7296.06 corers INR. In the post independent period India,s marine fish production increased from 0.5 million tonnes in 1950 to three million tonnes annually and foreign exchange worth ₹ 8,000 crore through seafood exports is being earned up to January 2009.

According to the annual report of Department of Animal Husbandry, Dairying and Fisheries Ministry of Agriculture, Government of India, New Delhi the State-wise Fish Production during the period 2006-07, depicts that four states with West Bengal on top of the list has produced the maximum of total fish products. According to the statistics West Bengal produced 1359.10 thousand tonnes of fish. Andhra Pradesh produced 856.93 thousand tones. Gujrat produced 747.33 thousand tones, whereas Kerala produced 677.63 thousand tones of fish.

Compared to growth in world fish production, fish production in India has increased at a faster rate mainly due to increasing volume of inland fish production (Krishnan, et al 2000). Inland fisheries and aquaculture has made rapid progress and are contributing around 50 per cent of the total fish production in the country (Krishnan *et al.*, 2001). By 2020, India, Latin America and China are projected to be the top exporters of aquaculture and fisheries resources.

India is indeed endowed with vast and varied fishery resource and a steady growth has been observed in harnessing fisher resources in the recent years. In the present decade least production was observed during 2000-01 when the total Fish production was 5.65 million tonnes including both Marine and Inland resources. Fish production during the year 2004-05 was 6.304 million tones comprising 2.778 million tonnes of marine fish and 3.526 million tonnes of inland fish. There has been steady growth in the export of fish Products. During 2004-05 the country exported 0.437 million tonnes of marine products, which resulted in export earning of ₹ 6188.92 crore (INR). Efforts are being made to boost the export potential through diversification of products for export. The country has now started exports of frozen squid, cuttle fish and variety of other finfishes. According to Department of Animal Husbandry, Dairying and Fisheries Ministry of Agriculture, Govt. of India, till 2004-05 about 6.74 lakh hectare of water area was brought under scientific fish farming through Fish Farmers Development Agencies (FFDA). This growth if not phenomenal is no doubt encouraging, which pushed India as the second important country in Aquaculture production next to China. The rate of growth in contribution of fisheries to Indias gross domestic product has started to approach the rate of growth in its gross domestic product.

Indian fisheries and aquaculture has proved to be an important sector not only for food production and providing nutritional security to the food basket but also for contributing to the agricultural exports and engaging about fourteen million people in different related activities. With diverse resources ranging from deep seas to lakes in the mountains and more than 10 per cent of the global biodiversity in terms of fish and shellfish species, the country has shown continuous and sustained increments in fish production since independence.

According to the reports of the National Fisheries Development Board, Government of India, during 2007-08 India contributed about 4.4 per cent of the global fish production. In India, fisheries sector contributed to 1.1 per cent of the GDP and 5.30 per cent of the agricultural GDP. Per capita fish availability had been 9.0 kgs and annual export earnings were 7,200 Crore (INR). 14 million employment opportunities have been generated by this sector till 2008-09.

3

Global Fishery Value Chains

The entire set of processes and activities which are required to produce and deliver a product to a target market is considered as supply chain. The term "produce" encompasses growing, transforming, or manufacturing. The entire chain goes from oceans or farms to hands, chopsticks and forks. Unfortunately, many central and local governments, donor agencies, NGOs are concerned with a subset of links within the value chain of fish and fishery products. Smooth functioning of value chain requires not only the factors of production and technology but also the efficient transport, market information systems and management. Supply chain flows from producer to final consumer in various ways and middle of the chain change the initial face of product and add incremental value to it.

This incremental value brings benefits to ends, producer and consumer. Especially, processing, distribution of products to different market places and other marketing activities need to cater for the consumer demands at right time and place.

VALUE CHAIN

The activities within the organization add value to the service and products that the organization produces, and all these activities should be run at optimum level if the organization is to gain any real competitive advantage. If they are run efficiently, the value obtained should exceed the costs of running them, *i.e.* customers should return to the organization and transact freely and willingly. Michael Porter suggested that the organization is split into "primary activities" and "support activities".

PRIMARY ACTIVITIES

Inbound logistics: Refers to goods being obtained from the organizations suppliers ready to be used for producing the end product.

Operations: The raw materials and goods obtained are manufactured into the final products. Value is added to the product at this stage as it moves through the production line.

Outbound logistics: Once the products have been manufactured, they are ready to be distributed to distribution centres, wholesalers, retailers or customers.

Marketing and sales: Marketing must make sure that the product is targeted towards the correct customer group. The marketing mix is used to establish an effective strategy; any competitive advantage is clearly communicated to the target group by the use of the promotional mix.

Services: After the product/service has been sold what support services does the organization have to offer. This may come in the form of after sales training, guarantees and warranties. With the above activities, and/or a combination of them maybe essential for the firm to develop the competitive advantage which Porter talks about in his book.

Support Activities

The support activities assist the primary activities in helping the organization achieve its competitive advantage. They include the following:

Procurement: This department must source raw materials for the organization and obtain the best price for doing so. For the price they must obtain the best possible quality.

Technology development: is the use of technology to obtain a competitive advantage within the organization. This is very important in today's technological-driven environment. Technology can be used in the production to reduce cost, thus add value, or in research and development to develop new products, or via the use of the internet so customers have access to online facilities.

Human resource management: The organization will have to recruit, train and develop the correct people for the organization in order to achieve its objectives. Staff will have to be motivated and well paid, in order to retain them for a longer period of time. Thus, the organization could get a competitive advantage through its motivated staff. Within the service sector, *e.g.* airlines it is the "staff" who may offer the competitive advantage that is needed within the field.

Firm infrastructure: Every organization needs to ensure that their finances, legal structure and management structure works efficiently and helps drive the organization forward. As you can see the value chain encompasses the whole organization and looks at how primary and support activities can work together effectively and efficiently to help gain the organization a superior competitive advantage.

Value Chain and Concepts of Marketing

Marketing concepts, practices and policies have largely influenced consumption behaviour. Marketing has influenced supply chain processes tremendously but at the same time hidden for most players in the value chain. In various industries supply chain processes have been driven by the marketing concepts developed. Major marketing management decisions can be classified in one of the following categories, Product, Price, Place and Promotion. These variables are known as 4Ps of marketing or marketing mix. Product is the physical product or service offered to the consumer. Whereas, price is the selling rate of the product and price decisions should take into account the profit margins and probable pricing strategies of competitors. Place or placement decisions are those associated with channels of distribution that serve as the means of getting the product to the target customers. The distribution system performs transactional, logistical and facilitating functions. Promotion decisions are those related to communicating and selling to potential consumers.

Traditional marketing model with 4Ps focused and doesn't translate well into the growing service-based economy we live in. Others criticize the marketing mix as a tool for setting marketing strategy because it does not have a goal. Criticism is the internal focus of the 4Ps, where the customer in the discussion of marketing, which however you dress it up, is concerned with persuading particular customers to buy. The traditional marketing mix is for suppliers pushing products into the marketplace and not customers pulling products out of the potential suppliers based on their needs and wants through market. If customers need to feel confident before they buy – to know, like and trust you – then there is no mention of the development of personal relationships which are an essential element of twenty-first century marketing. Present value chains focus more on services and they are composed of 7Ps. Services sector is growing fast and there are plenty of opportunities available for food service industry than the traditional food production Having the traditional marketing mix, the present value chain added: People — services are performed by people whose performance influences the quality of the service delivered and perceived to be delivered. Process – it is not just the attitudes of the person that matters but the process they use to provide the service. Physical evidence – services are intangible so that the customer looks for physical clues about the quality.

Value Chain versus Supply Chain

Value chains are concerned with what the market will pay for a product or service offered for sale. Moreover, market considerations differ from country to country, region to region and having close connection with food habits and consumption pattern of the people. The main objectives of value chain management are to maximize gross revenue and sustain it over time. Supply chains are concerned with what it costs and how long it takes to present the product for sale. The main objectives of supply chain management are to reduce the number of links and to reduce friction, such as bottlenecks, costs incurred, time to market etc. Good supply chain is essential to develop a value chain.

Fisher or shipper controlled and retailer controlled sections of value chain is explain. Moreover, fisher or shipper controlled value chains are cost driven while retailer controlled value chain are revenue driven. Key concerns of the producers are availability of fish in year round basis, minimise the seasonal gults and shortages and cater for service oriented customers with fresh produce. Retailer controlled value chains are more concern on value addition, differentiation, change the face of the product and focus more on private brands and labels. Especially, which facilitates the retail giants to cater for their brand loyal consumers and establish image in both local and international market.

THE EMERGENCE OF THE VALUE CHAINS IN THE FISH INDUSTRY

Similarities of fish marketing systems in developing and developed countries:

- Both have to face the same basic challenge of providing safe food having the right type and quality, to the right place and people. Right people refers to those who are willing and able to pay
- The market is composed of mixture of local and imported fish and fishery products
- Complex panorama of actors, enterprises and institutions
- Important role of supermarkets in fish and fishery product retailing
- Presence of hotels, restaurants and institutional channels indicating some food service suppliers
- Increasing role of regulations and standards.

Differences of fish marketing systems in developing versus developed countries:

- Vastly different scale at system and enterprise level
- Percentage of product handled formally lower in less developed countries
- Share of fresh vs. processed/manufactured fish much higher in less developed countries than emerging or developed countries
- Supermarket share is rising at a fast rate in developing countries to detriment of smaller retailers and wholesale markets
- Food service share and growth are becoming smaller because hotel, restaurant and institutional markets are less developed due to lower disposable income
- Standards less evolved and less complicated.

PESTLE ANALYSIS OF THE FISHERIES INDUSTRY

PESTLE analysis is a useful tool for understanding the "big picture" of the environment in which industry is operating, and environmental understanding will bring the advantage of the opportunities and guide to minimize the threats. PESTLE components are Political, Economic, Social, Technological, Legal, and Environment.

GLOBAL FISHERY VALUE CHAINS: A LITERATURE REVIEW

Fishery industry plays a significant role in the livelihood of more than 50 million people in terms of employment, income and provision of principal protein to the diet. Moreover, fisheries industry has ranked high among others in its contribution to the local and regional economy. The capability of fisheries industry to generate and sizable growth opportunities and to effectively contribute to the developing world's development objective of poverty eradication and wealth creation has been immensely disturbed due to constraints it focus.

Industry is at threatening levels due to overexploitation of resource base environmental degradation, climate change, high pressures on resources and poor or limited value addition. Value chains of pelagic fish in Asian developing countries are not developed to meet international market requirements and limited value addition. Main markets for pelagic species are domestic markets and processing efforts are poor. In contrast, value chains of dermasal species are well established and value addition generates profits to the stakeholders of the chain.

Face of the fisheries industry is affected by participation in either regional or global fish trade. Value chains are networks of labour and production processes where the result is a finished commodity. Value chains are led by firm leaders and chains consist of several nodes, each of which has a particular function in transforming an object from raw materials to an article of consumption.

Analysing finance in the fisheries value chain provides an interesting case because unlike for example grains, tree crops or vegetables, seasonality issues play less of a role. The case is also interesting for the complexity of interwoven value chains: fresh and processed fish, industrial and artisanal processing, domestic and export markets, food and feed products. Analysing the fishing value chain, with its unique social fabric and direct relationship to a fragile natural environment, also demands a discussion on the triple bottom line of economic, social and environmental issues.

TYPES OF GLOBAL FISHERY VALUE CHAINS

There are different fishery value chains in each country based on the number of actors involved in the process. Following figures represent the different value chains and some chains are totally operated and based on local markets and no international market interventions. Moreover, some value chains represent both local and foreign market interventions and make the market function more complicated.

FISH PRODUCTION

Capture fishery plays an important role of supplying fish to cater for consumer demands. To keep increasing fish supply, aquaculture is becoming an important occupation while bridging the gap between demand and supply. Supply chain of the marine capture fishery would comprise of several stakeholders, such as producer, wholesaler, dealers, middlemen, retailer, processor and consumer.

Fishermen and Fish Farmers

Capture fishery composed of two main actors, artisanal and industrial fishermen. Industrial fishers are concerned more about few economically important species and their scale of production is large compared with artisanal fishers. In Sinaloa, Mexico industrial fishermen produce 60 per cent of exported Sinaloa wild-caught shrimp, and artisanal fishermen produce 40 per cent. Five industrial producers dominate the market, providing them with leverage within the industry. Promarmex, the largest single producer coalition, contributes up to 70 per cent of Sinaloa's total wild-caught shrimp production. It is a vertically integrated actor in the value chain, which does its own processing, exporting, branding, and marketing. Conversely in Mexican shrimp fishery, between 4,000 and 5,000 artisanal fishermen organized into 140 local cooperatives aggregate shrimp for sale to the domestic and export markets. Each artisanal cooperative has a very small share of the market. Furthermore, artisanal fishermen are mostly restricted to bays and lagoons, in which they can only catch smaller shrimp, which are concurrently of less value in the market.

The value chain of the Nile perch, foreign hunter species found in Lake Victoria representing the establish aquaculture value chains in Africa. Formation of a Nile perch value chain from Lake Victoria to international markets was open up for new avenues to the local fishermen as well as other stakeholders of the chain. Length and number of nodes (participants) vary from country to country and from species to species. A study on dry fish value chain of Bangladesh identified four stakeholders: fish traders, wholesalers, medium operators between producer and consumer. The Hilsa market chain from fishermen to consumers encompasses mainly primary, secondary and retail markets, involving sales agents, suppliers, wholesalers and retailers. Backward stakeholders in both long and short supply chain were fishers, middlemen and wholesaler. Moreover, fish producers can be categorized into three groups, such as low income or poor of indebted producers, middle-income fishers and private entrepreneurs. Internal dynamics leads to create division between State Own Enterprises (SOEs) and local Chinese dominated private enterprises in Vietnam.

Fish quality as expected internationally guaranteed by the processing industry through adopting new quality guidance techniques and based on a strong selection of the quality of the fish at the receiving point. Vietnam's participation in value chains embedded in different institutional contexts has a profound impact on how local upgrading takes place. Common features of most developing country suppliers are insufficient quality control in the upstream part of the channel, insufficient use of ice and long-waiting times of the trucks. Moreover, in the upstream part of the channel, poor market information and incomplete information flows are common. Marginalization from global buyers is not necessarily a total exclusion as there is often third road to upgrading by for instance using socially embedded networks of say Chinese overseas to place products " through the back door" to the consumers. The main constraints at primary markets for fishermen are lack of bargaining power and market information.

Processors

Processors play significant role in international fishery value chains. Processors then sell their processed shrimp or fish to "Buyers" in the next segment of the value chain. This segment includes the retailers, wholesalers, and exporters and importers. Retailers buy relatively small quantities of fish or shrimp for local retailers and restaurants along the coast. All artisanal producers and some industrial producers pay third-party processors to process and pack shrimp, blue swimming crab before it is sold to forein buyers. Current processing standards make it difficult for United States importers to trace the products' origin and some processors lack technology and quality standards to meet the demands of large United States retailers. In general, traditional methods of fish processing (dry fish), and poor quality of products hinders the ways to enter into export market. Traditional processors are out of export market as they could not meet the Sanitary and Phyto-Sanitary measures and implications of Technical Barriers to Trade (TBT). Poor maintenance of quality standards hinders the progress of Vietnam seafood industry.

In Africa, Asia and Latin America, foreign investments entered the region and created a large processing capacity, but the benefits for the regions appeared to be limited because of a loss of traditional jobs and limited added value for the local population. Especially in Nile perch value chain, small-scale enterprises dominated the upper part of the value chain from fishermen to the processing industry.

The upper part of the chain was characterized by a lack of sufficient quality measures and control oligopolistic or monopolistic power in markets and incompleteness of market information. Especially in Vietnam SOEs have strong ties with local and central governments and their blessing receive capital to upgrading their systems to meet Hazard Analysis and Critical Control Point (HACCP) standard. Unfortunately, private local Chinese-based enterprises receive poor support to upgrade their systems and marginalized in HACCP standard certification process. Ethnic markets are playing an important role in establishing brand images and businesses in international markets. Especially in Vietnam, local Chinese-based enterprises use traditional technology, tastes and flavours to process fish and fishery products to the Chinese-dominated ethnic markets.

Some artisanal fishermen sell shrimp to intermediaries, who sell products to processing plants for the export market. Before being sold to buyers, shrimp are processed. Within the Mexican artisanal sector, processors tend to be third-party players who are paid to process shrimp, but are not directly connected to the sale of shrimp. Many large industrial fleets have their own processing plants, whereas smaller fleets or independent boats outsource processing much like artisanal fishermen. All processing plants are responsible for meeting and maintaining the quality and safety standards mandated by the Mexican government and the USFDA guidelines for imports.

Distributors and Traders

The next actors in the value chain are the distributors who store and sell products to retailers, food service and food management companies and restaurants. In general, fish catch is sold directly on the beach to various traders. Research has identified three types of distributors: speciality seafood distributors, full-line distributors and environmentally sustainable marketers. Speciality seafood distributors specialize in seafood products and develop regional supply chains.

Full-line distributors are selling a wide range of food products and they have national distribution networks. Dealers play key role in value chains and 60 per cent of the total catch is handled by them. Majority of dealers own vessels or make payment for the catch. Middlemen or commission agents handle remaining 40 per cent of the catch. Several stakeholders in fish handling contributed 22 per cent value addition to the raw fish. Exporters and importers represent Mexican export and United States import companies that buy shrimp to be sold internationally. Mexican Sinaloa shrimp is almost entirely sold through "Distributors," a segment that includes full-line distributors, speciality seafood distributors and environmentally sustainable marketers. Traders differ according to their scale of operation and in some countries based on ethnicity. Especially in Vietnam, vast majority of traders are local Chinese and often were former rice traders.

They have gained access to this lucrative market by networks based on trust that provide information on prices, the financial stability of buyers and access to credit. Traders usually handle large volumes of shrimp daily and need huge amount of capital. Credit is provided among traders and also between traders and seafood processing companies. SOEs are away from these functions and no long-term contacts with brokers and are not allowed to pay a higher price. Role of SOEs are very important to minimize the market failures and provide better prices for farmers.

One barrier to expanding the domestic shrimp market in Mexico is that coyotes or intermediaries play a very important role in bringing shrimp from the artisanal producers to the largest distribution markets. The intermediaries buy shrimp directly from artisanal fishermen and sell them to processing plants or to domestic operators. The benefit of these intermediaries is that they inject valuable liquidity into the artisanal sector, which enables artisanal fishermen to stay in business. The disadvantage is that artisanal producers earn less selling to coyotes compared with the domestic market value. Direct access to the market could enable fishermen to earn greater profits.

Retailers

Sales of domestic fish products in modern retail outlets, such as supermarkets, are limited in developing countries compared with developed country markets. The growing urban markets represent a market opportunity for fish farmers by improving their fish products through value-added processes, cold chain and linkages with supermarket segments. Grading and the use of ice are minimal in these domestic end-market channels, resulting in high spoilage levels. Retail chains secure as high as 150 per cent of profit. External dynamics, such as consolidation in both retailing and distribution in the main markets and Compliance on HACCP, create division between enterprises upgraded to HACCP system and non-HACCP enterprises.

Moreover, in Vietnam, local Chinese-based non-HACCP enterprises try to establish and distribute their brand names in small niche markets. SOEs and others with HACCP certification often maintain healthier ties with giant retailers. Regional networks enhance value-adding and close interaction between major buyers, such as Japan, EU, United States and Vietnamese suppliers in the chains. Most of the leading retail chains in Europe, United States, Australia, and others are catering to the growing ethnic markets with traditional tastes and flavours. A regional and global overseas Chinese seafood network based on Asian specialists provides space for branding and learning by doing on the export market. Products are then sold to "Retailers", including restaurants, grocery and speciality food stores, and food service and food management companies.

In the retailer and food service markets, the vast majority of shrimp purchased is farm-raised due to its lower prices and the high level of quality control buyers can have over the product. Mexican wild-caught shrimp processing and packaging could be upgraded to meet the demands and interests of large United States retailers. However, this step must be taken cautiously and with input from large retail buyers. Branding sustainable shrimp products is another opportunity to appeal to United States seafood distributors, retailers and food service companies, but their willingness to pay more for such a product varies. Only two small, cold-water sustainable shrimp products are available in the United States market. A product with MSC certification or one included on the Monterey Bay Aquarium list of recommended seafood products would likely be in high demand if it could meet the quality and packing criteria of retail buyers. Finding new market opportunities, such as these, may persuade fishermen to participate in new sustainability efforts. Marketing margins and profit for different species traded internationally varies widely. Hilsa value chain of Bangladesh reports lowest marketing margin and profit in primary markets and highest for secondary markets not in the retail markets.

Fish Consumption

Western Europe is the main fish consuming region among developed countries with per capita fish consumption of 22.2 kg. Europe has experienced a slow but steady increase in per capita fish consumption, which was 18 kg in 1970. Northern America (United States and Canada) and the developed Oceania (Australia and New Zealand) report per capita fish consumption similar to one in Western Europe of about 20 kg. For both regions, this represents a strong increase over the 1970s level of 14 kg. Among the main developed countries, Japan is by far the outstanding country with over 70 kg fish consumption, this is more or less the same per capita consumption than in the 1970s. Declining domestic catches were continually offset by higher imports.

Among developing countries, the lowest per capita fish consumption is reported by Near East countries which are as low as 4.8 kg. Africa, too, has only very limited fish consumption with per capita fish consumption of 8.4 kg in 1990, a certain increase from 7.9 kg recorded in 1985. In developing Asia (excluding China), the per capita fish consumption was 10.2 kg in 1990, up from 8.2 kg in 1970. This region aggregates countries with high fish consumption (Thailand, Indonesia) with countries having low fish consumption (India, Pakistan, Nepal). The growth of fish supply to China was the most outstanding event.

In 1970, the Chinese per capita fish consumption was only 3.6 kg, which increased to 9.9 kg in 1990. Countries having highest fish supply, such as Japan, Maldives and many islands, reported high levels of fish consumption. The highest rate was reported in the Maldives at 130 kg/year, in 1990, practically fish is the only food that does not have to be imported and is a stable food in this country. Iceland and Japan having around 70 kg per capita fish consumption are also among the countries with high fish consumption. In both countries, the fish consumption has stayed stable over the years. Many islands report high fish consumption which is not a surprise. The "Consumer" segment consists of the end consumers who purchase fish or shrimp from those in the "Retailers" segment. Environmental "NGOs" are working with actors across the value chain to reduce the ecological impact of fishing practices, purchasing decisions and consumption patterns.

Domestic Market

Shamsuddoha states that, number of intermediaries and stakeholders vary depending on the extent of the market. Localized market supply chain is too short. Standard common marketing chain exists in country's domestic market. Structure and functions of domestic fish markets differ from country to country. For example, there are three major commercialization channels in Mexico: (a) large distribution centres (centrales de abasto) in Mexico City, Guadalajara, Monterrey and most state capitals; (b) self-service stores and supermarkets; and (c) restaurants, hospitals, tourism and catering services. A fourth segment is the informal sector through which shrimp are sold in street markets, street cart, by which local government initiated a quota system for the first time in 2009 with the goal of developing a more responsible use of the fishery as a resource.

In general, fishes are traded in domestic markets in fresh, ungutted, whole and without adding ice. The travel duration between the primary markets and retail for urban markets is usually less than 12 hours. Moreover, if the transportation time is less than 6 hours from the primary market to the retail point, the fish is not iced; it is not done properly. Currently, there are no certification and regulation systems to improve the domestic seafood supply chain structure in Mexico. Facilities at domestic fish markets are minimal, with poor hygiene and sanitation and common among most developing country markets. There are no standard practices for handling, washing, sorting, grading, cleaning and icing of fish. In terms of volume, value and employment, the fish market in Bangladesh is large. Moreover, size of the Mexican domestic shrimp market is large: 80 per cent of total Mexican shrimp production is consumed domestically. Fish and fishery product consumption of production destinations are increasing sharply. For example, Mexican shrimp consumption grew at an average annual rate of 13 per cent from 2002 to 2008, increasing from 0.74 to 1.47 kg per capita. To satiate this increase in demand, Mexico imported 12,816 tonnes of shrimp in 2008, up to 30 per cent from 2007 (Téllez Castañeda, 2009). Growth in shrimp consumption is largely due to the increasing availability of inexpensive shrimp, which are mostly smaller, farmed products.

The fish marketing systems are traditional, complex and less competitive, but play a vital role in connecting the fishermen and consumers thus contributing significantly by in the value-adding process. A large number of people, many of whom live below the poverty line, find employment in coastal fish marketing as fishermen, assemblers, processors, traders, intermediary transporters and day labourers, including women and children. Before the institution of the international value chains, the local fishermen traded the fish they caught mostly to women (the fishmongers) on the communal beach or in the village or town market. The domestic chain is still characterized by small-scale fishermen and it has been marginalized by the export supply chain. The value chain describes the full range of activities that are required to bring a product or service from conception, through the different phases of production and delivery to final consumers. Facilities at fish markets are minimal, with poor hygiene and sanitation. There are no standard practices for handling, washing, sorting, grading, cleaning and icing of fish. The main constraints at primary markets are lack of bargaining power and market information. The marketing infrastructure, including cold storage, ice and transport facilities are generally inadequate, unhygienic and in disrepair. Comparatively, wholesale markets have better facilities, but in general conditions in primary and retail markets are far from satisfactory with regard to stalls, parking, speciality, sanitation, drainage and management.

International Market

The value addition is found highest by 105 per cent from wholesaler to retailer, followed by 90 per cent from wholesaler to exporter. Profit maximization and distribution are considerably high in short supply chains *i.e.* chains managed by private businesses entrepreneurs, NGOs and supermarkets. Traders follow different strategies to maintain healthier ties with producers and markets. For instance, in Vietnam, local Chinese brokers provide price guarantees and often up to 15 per cent excess price to farmers and this assures regular supplies. Moreover, companies without HACCP certification use their social networks to access both regional and international markets and enhance their power and coordinating skills in the value chain. They often develop new embedded networks in the local Chinese around the region.

In the case of Kenya, a wide variety of markets are linked to the capture fisheries value chain. The four main markets are the export markets for industrially processed fresh and frozen Nile perch filets, and the domestic markets for fresh tilapia, artisanally processed fish and feed grade omena. Additional set of markets are those related to Kenya's marine capture fisheries (shrimp, tuna, octopus, crab, etc.; Ardjosoediro and Neven, 2008). In the downstream part of the fish value chain, from processing industry to the export markets, information on prices, quality, quantity and standards is quite clear. Supply network of the international Nile perch value chain connects the small-scale upper part of the chain with the large-scale processing industry and international actors.

FINANCE IN THE VALUE CHAINS

Fisher's knowledge and experience on finance and management is poor and hinders the success of the industry. Only the industrial fishermen focus more attention on finance function and other management aspects and earn healthier profits. Common features among developing country fishers are enormous cash flows, low financial literacy levels, low saving culture, largely operate outside the formal financial systems, weak financial functions, heavily depend on informal financial sources that are unreliable, inadequate and highly expensive, poor business management skills, weak community organization with high levels of political interventions and vertical power imbalances.

BOTTLENECKS OF THE GLOBAL FISHERY VALUE CHAINS

The formation of the international fish value chain created new market opportunities that stimulated local economy. However, investments in local infrastructure or human resource development were low despite of the growth in trade. Traditional jobs such as female fishmongers, mobile fish traders, small-scale local processors etc. were lost (Abila and Jansen, 1996) and insecure jobs in processing factories. The added value for the local population has been low and always premium quality prepared for export. In Mexico, first-grade wild-caught shrimp are primarily exported and lower-grade wild-caught shrimp are sold in domestic market. Wild-caught shrimp is expensive in Mexico; thus, the domestic market opportunities are more limited.

Fish is an important and main protein supplement to the local people's diet where other protein sources are scarce. Malnutrition and especially protein malnutrition is common among most of the fish exporting developing nations. High demand for fish and fishery products and developed country markets are ready to pay for the premium prices lead to worsen the situation. Only the low-quality fish and juveniles, which are not accepted by processing industries, are left for local consumption actors.

Export processing industry focuses on few species and implies considerable risk on fish stocks and biodiversity. On one hand, some species are economically important, such as tuna for Maldives, Thailand and Indonesia; shrimp and prawn for Thailand, Bangladesh and Mexico; Nile perch for Kenya, Morocco, Nigeria etc. On the other hand, some species are socially important and national food habits and culture closely attached to that species, such as Hilsa for Bangladesh, Tilapia for Thailand, Kenya China and etc. Moreover, some species are both economically and socially important for respective producer destinations. In general, priority has been given to the exports and remaining low-quality fish comes to the domestic market.

Moreover, plight of small-scale subsistence fishers were lead to develop social unrests. There were no incentives for fishermen to change the current *status quo*. Small-scale fishermen are highly dependent on buying agents and become victim of unique power distribution. Local actors, such as small-scale fishermen, processing industry, credit institutions and public sector, are heavily dependent on other actors of supply chain. But chain has provided new opportunities. The international actors, such as traders, retailers and retail chains and customers, are especially interested in price, convenience and healthiness of the product. Their concern on biodiversity, environmental sustainability and welfare of fishing communities are low. Both local and international actors often look only after the part of the chain, in which they operate to make a profit. Moreover, lack of coordination between local and international actors is common and makes an appeal to have a joint action or supply chain governance system.

Quality assurance and certification are key challenges to the developing country suppliers, where most of the producers are far from the international market requirements. Poor-quality assurance and certification systems lead to lowering their profit margins. In general, they act as raw material suppliers to the developed country industrial processors. A future challenge for the local Vietnamese Chinese processors participating in regional and overseas Chinese networks is to upgrade their systems to HACCP. The regional markets will most likely require stricter food safety standards and rising numbers of regional and global retails which mainly consider certified products.

Over fishing, illegal and unregulated fishing, stock depletion and environmental population are key threatening issues to the fish producers around the world. Stocks of most of the economically important species are now reached up to alarming levels (more than the maximum sustainable yields) and authorities have to set a change in the nature of fishery from open access to private ownership property. There are clear indications that Nile perch has reached its maximum sustainable yield, and overfishing in Kenya's shallow lake waters is now reducing the landed volumes and Fisher folk are forced to incur higher costs to go to deeper waters for decreasing quantities of fish caught per trip. Although there are associations and all the beaches have beach management units as part of the government fisheries policy, there is little collaboration between fisher folk in terms of procurement, fishing or marketing. One challenge the fishery faces is tracking the large number of artisanal and commercial vessels. The goals of the government are to improve fishery management by assigning permits to restrict access, regionalizing fishing efforts, identifying legal fishermen, cooperatives landing sites, and improving social and economic conditions for fishermen. Boats registered by the government have a non-replicable microchip, which helps to identify each cooperative, permit and landing site. The microchip is fundamental to improve capture registration and establish a relationship among the boats' production, port and coastal system.

Different institutional contexts of end-markets are linked to different forms of coordination and control of global value chains. Economically and socially important species and value chains differ widely across regions and globally. Both local and regional networks enhance the value addition, brand creation and brand strengthening, technology enhancements, profitability and market access. Value chains of the most economically important species and destinations are needed to develop vision on; learning investment, market access, sales, and exports. Smooth functioning of value chains need to assure the favourable policy environment as well as good governance system. There is an important need to identify and support promising value chains with assistance at key point in the supply chain based on collaborative analysis of challenges, joint definition of priorities, and expert assistance from industry-experienced people.

Take a cluster approach only as the starting point for value chains, not as an end in itself. All actors or stakeholders of the value chains should concentrate on competitiveness and productivity and look for and exploit multiple ways to add value once initial success has been attained with a single deal. Ensure sustainability within the value chains is key important feature to cater the changing demands. An important need recognize that some keys to success require mainly public sector intervention, others only private, and some are a mixture of the two. Moreover, fisheries industry need to seek private sector alliances at all stages of supply and value chains for better future.

Strengthening the weak financial structure, focus more on formal financial systems, reducing power imbalances in the governance structures and low political intervention in community level organizations, and resolving socio-cultural and environmental concerns are the major concerns on development of value chains in developing countries. The high levels of post-catch losses indicate that the urgent need of an introduction of coolers and improved ice distribution systems, proper harbours, landing sites and markets would be an upgrade strategy that could stimulate value chain growth. While this could indeed lead to higher profitability at first, without retaining these profits and reinvesting them back into their business, value chain actors will not be able to grow their business. Good governance systems, protection of remaining stocks and stock enhancements, stop illegal and unregulated fishing practices, improve welfare of the fishing communities, mitigation measures to climate change, etc. are the crying needs of the hour. This risk needs to be addressed through a systemic enforcement of environmental protection measures and a diversification strategy.

THE PHILIPPINE SEAFOOD SECTOR: A VALUE CHAIN ANALYSIS

The Asian region is a major supplier of fish products to the EU market. Over the period 2005-2010 in particular, the aquaculture sector in some Asian countries became an important producer as well as exporter of whitefish and shrimps, according to information compiled for the Centre for the Promotion of Imports from developing countries (CBI) by LEI, part of Wageningen UR.

Within the Asian region CBI is currently studying the possibilities of developing in-tegrated programmes for the seafood sector for specific countries. This follows up on CBI's current sea-food activities in Indonesia with the Ministry of Marine Affairs and Fisheries (MMAF) and the Surabaya Seafood Centre.

Based on the results of the desk study which was carried out in phase one of this seafood export VCA, the following subsectors in the Philippines were selected for value chain analysis:

- Shrimp
- Tuna
- Seaweed

SHRIMP SUBSECTOR

The Philippines used to be one of the prime movers in the Asian shrimp industry. At present this is no longer the case. Production of Black Tiger shrimp was only 10,000 tonnes in 2010 due to disease out-breaks and crop failures, which resulted from lack of quality seeds and bad farm management practices. The production of Pacific White shrimp is currently limited to the domestic market that is estimated to be around 4,500 tonnes a year.

Three main bottlenecks for the export potential of the Philippine shrimp sub-sector have been identified as a result of the desk study, field work and validation workshop.

Table. Main Bottlenecks of the Shrimp Subsector

Bottlenecks	Level in the value chain
Lack of EU certified processing establishment	Processors and exporters
Lack of competitivenessof white shrimp	Primary production
Traceability	Primary production and trading

Shrimp exports are confronted with significant problems from the domestic supply chain as well as from the international market. Participants during the conference expressed their hope that if more companies get EU certified, the sector would receive a boost. However, this is not at all certain, as many export companies who previously were EU certified have not renewed their EU approval because the costs were higher than the expected benefits.

The companies that are trying to get approved now face difficulties with complying with the BFAR procedures. Despite the current crisis, there is great potential in the Philippines for shrimp farming. Conference participants underlined that EU buyers often are not aware that shrimp is produced in the Philippines.

The two companies that are EU certified both found customers in the EU and export Black Tiger shrimp. These two companies are the largest shrimp exporters in the country. If other smaller or medium-sized companies get EU approval they will need additional support to get market infor-mation and to make themselves visible in the EU market.

SEAWEED SUBSECTOR

More than 70 per cent of total seaweed production comes from Mindanao, in the southern part of the country. In addition to the local production, the Philippines also imports Raw Dried Seaweed (RDS) to fulfil the demand from local carrageenan processing companies. The combined volume of RDS from local production and imports was more than 90,000 tonnes in 2010. Four main bottlenecks for the export potential of the Philippine seaweed subsector have been identified as a result of the desk study, field work and validation workshop.

Table. Main Bottlenecks of the Seaweed Subsector

Bottlenecks	Level in the Value Chain
Lack of finance and investment	Primary production
Lack of government support (R&D)	All levels
Limited market access to export markets for carrageenan	Processors and exporters
Strong competition from international RDS buyers and cheaper gums	Processors and exporters

The Philippine seaweed sector is strong and offers employment to lots of coastal communities and also contributes to foreign trade. The Philippine carrageenan processing industry is the second strongest in Asia after China. However, it is under threat of competition from China and, increasingly, Indonesia. In order to survive, it is crucial for the sector to increase domestic production and productivity. Improved productivity could reduce production costs and make the product more competitive. It is clear that to achieve this, the government and private sector need to invest both capital and knowledge in the seaweed farming sector. Also, the sector should organise itself better and overcome disputes about competition that hamper cooperation between companies and suppliers.

Although it is often argued that carrageenan exporters should be able to survive on their own because they are full grown businesses which are often in the hands of multinational companies, there is also a group of local business which are struggling to survive. Some of these have already left the sector and others look for inventive strategies and product development to maintain their position.

This group of companies is in need of support to increase their visibility in the international market and to get market intelligence about opportunities for marketing innovative products such as organic seaweed fertilizer. The issue here is that the products produced by these companies are not suitable for seafood trade fairs. Therefore it is doubtful whether the seaweed sector fits in the seafood programme. The Philippine government has made the same conclusion by placing the carrageenan exporters under the authority of DTI instead of BFAR.

FROZEN FISH SUBSECTOR

Less than 3 per cent of total fish production in Bangladesh is exported. Exported species are a variety of captured fish of which the most well known is the hilsha shad (Hilsha ilisha). The cultured species with the highest export potential are pangasius (Pangasius Hypophthalmus) and tilapia (Oreochromis spp.). In 2010 the total value of fish exports was USD80m, of which frozen fish contributed almost USD35m. The most important markets for frozen fish are the UK, Saudi Arabia, the US and to some extent Italy and China. Most products are exported as block frozen and an almost negligible part as fillets. Five main bottlenecks for the export potential of the Bangladeshi frozen fish subsector have been identified.

Table. Summary of Bottlenecks in the Bangladeshi Frozen Fish Subsector

Bottleneck	Level in the Supply Chain
Lack of skilled labour fource	Processing
Dominant position of traders and commission agents	Supply chain
Lack of cold storage facilities	Supply chain
Lack of supply	Primary production
Low quality of the fish	Primary production

Although exports of whole frozen captured fishes still have a good potential in the countries that have Bangladeshi expat communities, this study focused on the export potential of cultured fish because aquaculture species have the highest development potential. The cultured species with the highest export potential are pangasius and tilapia. However, pangasius and tilapia are also consumed locally and therefore yield high prices in the domestic market. As a result most pangasius and tilapia are consumed locally and exports are limited. At this moment it is not expected that export promotion activities would immediately increase the export volume or value of Bangladeshi frozen fish because the problems the subsector is confronted with at the level of primary production and other levels of the supply chain are too large.

The lack of supply is mainly caused by low productivity of fish farms, post-harvest losses and a strong local demand for fish.

There are three strategies suggested to deal with the lack of supply:

1. Increase the productivity of fish farms,
2. Reduce post-harvest losses and
3. Encourage exporters to invest in integrated fish farms.

Strategy one should include training programmes for farmers that support them to reduce mortality rates and increase the productivity of the ponds. Strategy two should include investments in the infrastructure including cold storage facilities and proper transport. Strategy three should enable exporters to generate a constant supply of fish that does not reduce the availability of fish on the local market and of which the quality is ensured because the exporter can control all the inputs and invest in more intensive and better managed production systems. This would also make the exporters less dependent on the middlemen who dominate the supply chain and often are not quality minded.

Low quality of the fish partly is due to a lack of proper inputs and poor farm management but also due to a lack of cold storage facilities along the supply chain. Low-quality inputs and poor farm management result in fish that have no white but a yellowish or reddish colour of the fillet. These fillets are not suitable for many export markets. A lack of cold storage facilities threatens the freshness of the products that reach the processing establishments. To deal with the low quality of the fish that reaches the exporters, strategies should focus on (1) improving the quality of inputs (feed and seed) and (2) improving the infrastructure along the supply chain. The second strategy includes increasing the availability of cold storage facilities (the fourth bottleneck).

The final bottleneck, the lack of skilled labour force, limits the processing capacity of exporters. Currently, most frozen fish is exported as block frozen items while the highest market value and demand are for fillets. One explanation is that fish exports are often regarded as a sideline and an option to use the processing capacity more efficiently when there is a low supply of shrimp, which is mostly the main export item. Consequently, fish processing techniques receive insufficient attention from managers and factory owners. A strategy that focuses on creating a force of skilled factory workers who have the skills to properly handle fish fillets would contribute to the potential of Bangladeshi fish fillets in the international market.

It is expected that improved productivity, quality, and increased processing skills have a positive impact on the price and competitiveness of Bangladeshi frozen fish in the international market. However, the important question remains to what extent the promotion of export of frozen fish conflicts with the provision of local Bangladeshi food security because for Bangladeshi consumers fish is the most important source of animal protein.

RECENT TRENDS IN FISHERIES PRODUCTION

THE REGIONAL PERSPECTIVE

Asia

Asia is the world's foremost capture fishery and aquaculture producer. In 2004 the regional marine harvest alone totaled 32 million tonnes of marine fish, 16 million tonnes of molluscs, 14.4 million tonnes of aquatic plants and 5.9 million tonnes of crustaceans. Although the growth in marine fish landings since the early 1990s has been rather modest (up 23 per cent from around 26 million tonnes) due to the gradual (over) exploitation of stocks, crustacean extraction has doubled and the harvesting of molluscs and aquatic plants has trebled over the same period. The principal marine fishing nation in the region is China, although a further nine states landed in excess of 500 000 tonnes in 2004. The Table, nevertheless, fails to illustrate the profound re-alignment in regional production that has taken place since 1990. In that year Japan landed 8.1 million tonnes (30.7 per cent of regional landings) and the Republic of Korea 1.8 million tonnes (6.8 per cent). However, a series of fishing capacity management plans introduced in the two countries during the intervening years has seen capacity—and with it marine landings—sharply reduce in both countries. In contrast, production has soared in most of the less developed economies (Chinese landings up 147 per cent, Vietnamese up 132 per cent, Indonesian +92 per cent, Myanmar +84 per cent), India being somewhat of an exception (production up 23 per cent) due to its principal marine fisheries already being rather closer to full exploitation.

The principal molluscs extracted are the high-value Pacific cupped oyster-*Crassostrea gigas* (4.3 million tonnes in 2004) and Yesso scallop-*Patinopecten yessoensis* (1.5 million tonnes), and the lower value Japanese carpet shell-*Ruditapes philippinarum* (2.9 million tonnes). While China dominates mollusc production in the region (61 per cent of reported 2004 regional harvest), Japan (1.8 million tonnes, 11 per cent of regional harvest), the Philippines (1.3 million tonnes, 8 per cent) and the Republic of Korea (800 000 tonnes, 5 per cent) also possess significant mollusc-extraction sectors. The main aquatic plant harvested is Japanese kelp, the harvest of which doubled to just over 4.6 million tonnes during the period 1990-2004, although regional production of *Wakame* (a kelp-like plant), laver (*Nori*) and Zanzibar weed have also surpassed the million tonne mark (2.5 million, 1.4 million and 1.1 million tonnes respectively in 2004). The main driver in aquatic plant extraction is, once again, China (76 per cent of 2004 regional harvest). Although prawns/shrimp dominate under the crustacean category (73 per cent of category volume in 2004), serious viral pathogens affecting the region's main indigenous *penaeids* (the Giant tiger prawn and the fleshy prawn) constrained growth for much of the 1990s. The introduction of the whiteleg shrimp at the turn of the century however promptly restored momentum to the sector, with this one species alone providing a harvest of 710 465 tonnes (17 per cent of regional shrimp/prawn harvest) in 2004. China supplied 53.5 per cent of the regional crustacean catch the same year.

Regional production from inland water fisheries has grown more than three-fold since 1990 to 27.4 million tonnes in 2004, although the majority of this (86.4 per cent) arises through aquacultural operations. China dominates once more (67.8 per cent of 2004 harvest), having increased landings from 4.9 to 18.6 million tonnes over the same period. India (2.95 million tonnes), Bangladesh (1.5 million tonnes), Indonesia and Myanmar (824 000 tonnes apiece) and Thailand (531 000 tonnes) also post significant inland landings. Carp—chiefly Common, Crucian, Grass and Silver—account for two-thirds of the regional inland fisheries output.

FAO calculations suggest that aquaculture production provided an income in 2004 of around US$27.9 billion for participants, up from US$7.3 billion in 1990—with the main beneficiaries currently being China (US$23 billion) and India (US$2.2 billion). Per caput food supply varies across the region being highest in East and Southeast Asia (26.8 kg), and markedly lower in both Southern Asia and the Near East (both 5.6 kg). Per caput supply is particularly high in Japan (64.7 kg), where the tradition of eating fish is very strong and fish is generally more important than meat in the diet, while the converse holds true in Afghanistan (zero kg). In Bangladesh, Indonesia, and Sri Lanka fish provides more than half of the daily animal protein requirements.

Europe

In Europe (including the former USSR and the new transition economies) marine catches have shown a sharp decline from the 20 million tonnes that was regularly landed in the 1980s, with the most recent harvest data (2004—12.6 million tonnes) being on a par with the quantities harvested in the early 1960s. The major contributor to this decline has been the collapse of the USSR—with Soviet landings dropping from 11 million tonnes in 1987 to just 2.5 million tonnes in 2004.

In Western Europe, the intensive exploitation of many demersal groundfish stocks (cod and haddock, for instance) during recent decades has meant that some stocks are now considered to be outside the safe biological limits and, as a consequence, European Union (EU) fleets are increasingly seeking access to other countries' exclusive economic zones (EEZs) for their distant-water fleets. The major (over 0.5 million tonnes) European marine fishing nations in 2004 were the Russian Federation, Norway, Iceland, Denmark, Spain, the Faeroe Islands and the United Kingdom.

A similar trend is evident in inland waters. Here catches have fallen from a late 1980s peak of 450 000 tonnes to 206 062 tonnes in 2004, the principal factor in the decline being a sharp fall in reported landings from the ex-Soviet economies.

The one success story has been aquaculture. In 2004, 2.24 million tonnes were produced at a value exceeding US$5.58 billion—up over 50 per cent in value and volume terms since 1994—with farmed diadromous species (principally salmon and trout) accounting for the bulk of the growth and the major portion of current worth (57.5 per cent of 2004 revenues). The main producing countries were Norway and, to a lesser extent, Scotland.

While current fish per caput supply levels in the region as a whole (at around 17 kg) are slightly above the global average, levels vary considerably among sub-regions and countries. While in the industrialized countries the average per caput supply is about 22.5 kg per year, in Iceland mean fish supply exceeds 90 kg per capita. In contrast, in Bulgaria, Rumania and Serbia-Montenegro, annual supply is less than 4 kg per capita. Latin America and the Caribbean. In Latin America and the Caribbean, total marine landings are heavily influenced by fluctuations in small pelagic (principally anchoveta) stocks, and periodic stock collapses (such as the anchoveta collapses in 1973 and, to a lesser extent, in the early 1980s—and subsequently in 1998) result in sharply reduced harvests.

This has been slightly offset by an increase in demersal production (particularly hake) and squid in the southwest Atlantic and tuna across the region, but most major demersal and small pelagic stocks are now considered to be fully-or over-exploited. Marine production in the region peaked in 1994 when 22.8 million tonnes was landed, with current (2004) landings around 17.5 million tonnes. Major Latin American marine fishing nations (over 0.5 million tonnes) in 2004 were Peru, Chile, Mexico and Argentina.

In comparison, Latin American inland fisheries are rather less significant, yielding around 325 000 to 375 000 tonnes per annum over the last few decades, with the recent rise in recorded landings (to around 440 000 tonnes) more due to better recording of catches than the exploitation of new fisheries. The same cannot be said of aquaculture, however. The abundance of small pelagic fish provided the stimulus for the development of an important fish reduction industry (currently consuming more than two-thirds of the total marine catch) and with it, a domestic aquaculture industry. The contribution of aquaculture—in volume and value terms—increased by more than threefold over the decade to 2004, production increasing from 0.34 to 1.14 million tonnes while the value of aquacultural output grew from US$1.32 to 4.56 billion. The main impulse for this growth came from the expansion of freshwater fish and marine diadromous (essentially salmon and trout) farming—up five and six-fold respectively, with Chile exclusively responsible for the upsurge in diadromous output and Brazil largely accounting for the increase in freshwater fish landings. As in other regions, fish supply and consumption levels vary widely. In some of the small Caribbean island states (Grenada, Montserrat, the Turks and Caicos islands) per caput supply exceeds 40 kg p.a. —substantially higher than in either Central America (around 9.3 kg p.a.) or South America (8.8 kg p.a.)—while its contribution to daily animal protein requirements ranges from 1.7 per cent in Bolivia to 60.5 per cent in Belize.

Africa

African marine catches, having risen from 3 million tonnes in 1990, stabilised at around 4.8 million tonnes at the start of the twenty-first century. However, these aggregate figures mask a number of differing trends. Landings in North Africa, which accounted for around 26 per cent of the 2004 catch, are dominated by Morocco (67 per cent of sub-regional landings), which realizes 95 per cent of its catch in the Atlantic. Contemporary evidence suggests there is little scope for further exploitation of Mediterranean fish stocks. Southern

Africa harvests have grown above trend, enabling the region to increase its share of continental landings from 30 to 34 per cent over the period 1990-2004.

Production is dominated by South Africa (52 per cent of sub-regional landings), with its focus on the Cape hake fisheries, and Namibia (34 per cent), where the emphasis is on small pelagic stocks. Eastern sea-board landings are much less important, accounting for under 4 per cent of continental landings in 2004, a percentage that has shrunk moreover as landings in the sub-region have risen at well below trend rates since 1990. The Western sea-board supplied 36 per cent of the marine catch in 2004, largely drawn from Senegal (374 245 tonnes in 2004), Ghana (313 935 tonnes) and Nigeria (251 232 tonnes). Although the Ghanaian catch in 2004 is of a similar magnitude to that recorded in 1990, Senegalese pelagic landings have increased dramatically (100 000+ tonnes) over the same time-span.

Africa has the world's most important inland fisheries, 2.25 million tonnes being harvested in 2004 compared to 1.8 million tonnes a decade previously. Here too though the aggregate figure obscures regional variations. Tanzania and Kenya, who source a large portion of their inland catch from the Lake Victoria Nile perch fishery have experienced landing declines of 16 and 38 per cent respectively since 1990 (although Ugandan catches have remained rather more buoyant).

Elsewhere, however, the story has been one of increased exploitation of inland water resources—catch estimates for the Democratic Republic of Congo suggest landings now stand at 217 000 tonnes (up 35.6 per cent since 1990), Egypt has seen output leap by 52 per cent (to 238,455 tonnes), while Nigerian landings have doubled to 215 000 tonnes.

Aquaculture, by contrast, is less well-developed in the region, output reaching 570 113 tonnes, worth US$893 million in 2004. Nevertheless, this represents a substantive improvement when set beside the equivalent 1990 figures of 82 475 tonnes and US$166 million.

Egypt is the principal aquaculture producer, presently accounting for around 83 per cent of volume and 69 per cent of value, with common carp culture well to the fore. Other countries with briskly growing aquaculture sectors include Kenya, Madagascar, Nigeria, South Africa and Zambia, all of whom have doubled their annual production (admittedly, from a small initial base) several times over the last decade or so.

Per caput fish supply in the region is low by international standards. Sub-regional averages lie between 3.7 kg (East Africa) and 11.6 kg (West Africa) p.a., with Niger (1.6 kg) and Gambia (29.3 kg) representing the extremes. Fish, historically, has nonetheless been a popular food item, providing more than half an individual's daily animal protein needs in Gambia, Ghana and Sierra Leone and—since most parts of the fish are consumed—it has also contributed significantly to calcium and iodine intakes.

However, the growth of the fish trade with the EU (facilitated, in part, by fishing agreements between the EU and various African nation states), Japan and, more recently, China, is leading to a reduction in the domestic availability of fish in both the main West African nations and those landlocked Central African countries that used to import fish from West Africa.

North America

In North America, marine catches declined steadily from a peak of 6.8 million tonnes in 1988 to 5.6 million tonnes in 2004, principally due to the overexploitation of the main commercial groundfish stocks in some areas where fisheries are now closed or subject to restrictions.

The most notable closure was in 1992 when Canada closed its northern cod fisheries off the Grand Banks, and over 20,000 people lost their jobs. Canadian landings, as a consequence, have fallen rather more sharply (down just over 50 per cent since 1990, to 614 107 tonnes) than US landings over the last decade or so.

A similar trend is evident in inland waters, the take dropping from a 1989 peak of 205 503 tonnes to the current level of 143 370 tonnes (down 30 per cent), despite the growing popularity of sport fishing, most notably in USA where federal agencies have been particularly active in developing recreational fisheries programmes since 1995. In the aquaculture sector, however, production in both volume and value terms has more than doubled since the early 1990s (currently output is 955 178 tonnes worth US$1.99 billion).

Aquaculture in North America—unlike in Europe or South America—is a much more diversified industry which includes marine, freshwater and diadromous fishes, crustaceans, molluscs and plants, although salmon and trout dominate in Canada, while catfish is an important sub-component of the US industry. Per caput fish supply levels in North America—at 22.7 kg per head—are comparable to the levels recorded in the industrialized countries of Europe, and sharply higher than the rates posted in South and Central America.

Oceania

Oceania provides comparatively little of the global marine catch (<2 per cent), landing 1.4 million tonnes in 2004 although this is certainly an underestimate as much of the coastal catch in the region's small island developing states (SIDS) goes unreported. Although catches have risen by around 53 per cent over the past decade, concerns over the current status of major commercial stocks has prompted the introduction of individual transferable quotas (ITQs) in New Zealand, the main regional fishery. Besides New Zealand (44 per cent of regional landings—cod and molluscs each accounting for a third of the NZ catch), the other major regional fishing nations are Australia (18 per cent, a quarter of which comes from the crustacean catch) and Papua New Guinea (15 per cent, almost exclusively tuna). However, for many SIDS it is a key strategic sector in the quest for development given its contribution to production, local food security, employment and export earnings. Landings from inland waters are equally low in global terms, with a reported catch of just 11,313 tonnes in 2004—largely from Papua New Guinea (96 per cent of total), where inland fisheries in the highland areas are an important source of food security.

However, regional data does not presently include sport and recreational fisheries—important fisheries in both New Zealand and Australia. Oceania's contribution to global aquaculture output is also relatively small, amounting to 139,273 tonnes worth US$447 million in 2004, although it has grown by 86 per cent and 166 per cent in volume and value terms respectively over the past decade.

Production is almost exclusively concentrated in New Zealand (66 per cent of volume, 37 per cent of value) and Australia (28 per cent of volume, 59 per cent of value) and, unlike in other regions, it is mollusc cultivation (principally mussels in New Zealand and oysters in Australia) rather than diadromous fish farming that dominates.

Fish are an important protein source for many Pacific islanders—supplying as much as 77 per cent of the Solomon Islands' and 85 per cent of the Maldives' animal protein needs—although culturally and nutritionally it is of lesser importance in both New Zealand and Australia. While under-reporting tends to bias supply figures downwards for many of the Oceanic countries, nevertheless 2004 per caput supply ascended to 185.9 kg in the Maldives (the highest in the world), 75.2 kg in Kiribati and 94.3 kg in Palau.

SUMMARY

The value chain analysis starts from the input suppliers to final buyers and the relationships among them. It analyses the factors in-fluencing industry performance, including access to and the requirements of end-users; the legal, regulatory and policy environment; coordination between firms in the industry; and the level and quality of support services. Relationships among firms in an industry can facilitate production and marketing efficiencies and enable the flow of information, learning, resources and benefits. This chapter aims to present the major investigations of the global fishery value chains.

Value chain analysis in a fishery is a powerful tool for all stakeholders, from fishers to fishery business operators to identify the key activities within the industry which form the value chain for that industry and have the potential of a sustainable competitive advantage for an industry. There in, competitive advantage of fishery lies in its ability to perform crucial activities along the value chain better than its competitors. First chapter of the report aims to identify and discuss the concept of value chain, purposes and importance of value chain. Chapter two and three was aimed to identify and discuss the drivers and governors on change of demand and supply. Fourth chapter focuses on global fishery value chains, different models and theoretical underpinning for the value chain analysis. In addition, value chain literature and PESTLE analysis received weighted discussion.

4

Fishery Management: Authorities and Policies

Wisconsin State Statutes 15.34 established the Wisconsin Department of Natural Resources (WDNR) and Statute 23.09 provides the general authorities and limitations for fisheries management and regulation within the State. In addition, the WDNR's Administrative Code further defines the agency's responsibilities to administer fisheries management programmes based on scientific management principles that emphasize the protection, perpetuation, development and use of all desirable aquatic species. Aquatic resources include both non-game and game species of fish, other aquatic animals and their habitats. Endangered and threatened species form a special group that are managed pursuant to additional State statutes. The goal of fish management is to provide for the optimal use and enjoyment of Wisconsin's aquatic resources, both sport and commercial, while maintaining a healthy and diverse environment. Management of fish and other natural resources is based upon a trust doctrine which secures the right of all Wisconsin citizens to quality, non-polluted waters and holds these waters as common property for all citizens. Fish management programmes vigorously uphold the doctrine that citizens have a right to use, in common, the waters of the state and that these waters shall be maintained free of pollution.

To assure its effectiveness, the fisheries management programme maintains close working relationships with all functions of the department, other governmental agencies, federally recognized Native American tribes, and the public. The WDNR keeps interested parties informed of policies, plans and management actions. To anticipate change and meet future demand, the department engages in longrange management planning. A comprehensive overview of the historical evolution of Wisconsin fishery management policy is outlined by Puff. The overriding vision for Wisconsin's fishery programme was redefined and updated in Fish Wisconsin 2000. Fish 2000 is an internal policy document designed to guide fisheries management into the next century. It calls for expanded partnerships and reprioritizing the management focus. The new vision calls for balancing coldwater and warmwater management and in focusing management on entire communities or ecosystems rather than single species. This sciencebased resources management policy applies ecosystem management principles and practices, builds partnerships with other agencies and landowners, and develops innovative and proactive information and education strategies regarding biodiversity and its relationship to ecosystem management.

RECREATIONAL FISHING

The WDNR regulates the sport and commercial harvest of aquatic resources to achieve optimal sustained yields. Sport fishing is closely regulated to provide all State residents an equal opportunity to safely enjoy the aquatic resources. To meet this public trust responsibility, the WDNR protects and enhances fish and other aquatic resources, regulates fishing effort to ensure that it does not exceed the capabilities of the resource to sustain desirable, quality fish populations, and integrates social and economic values into fisheries management programmes and actions. Emphasis is placed on fostering a sense of responsibility for the resource in all who participate and enjoy fishing, minimizing user conflicts and ensuring that aesthetic and cultural values associated with fishing are held in trust for future generations.

FISH STOCKING

Wisconsin State Statute 23.09 provides the WDNR the authority to capture, propagate, transport, sell or exchange any species of game or fish needed for stocking or restocking waters of the state. Such actions will be specifically designed and carried out to meet fisheries management objectives and provide for optimal sustained yields of sport and commercial fisheries. The WDNR is authorized and responsible for the propagation, rearing and distribution of fish for stocking in waters lacking adequate natural reproduction and where reasonable returns of stocked fish are demonstrated by surveys. Stocking priorities are based on use opportunities, hatchery production capabilities, cost and habitat potential. The State operates 14 hatcheries to meet the demands of fish for restoration and enhancement stockings.

Exotic Fish Management

The department may, where feasible, control fish populations that are stunted or harmful to more desirable fish species. Control measures may include mechanical removal, predator stocking, commercial harvest and chemical treatment. The stocking of exotic species for biological control purposes (*e.g.*, grass carp), or to create and/or enhance sport-fishing opportunities, is thoroughly evaluated for ecological risks, cost-benefits, and potential effectiveness in meeting management objectives. Programmes to monitor and control nuisance non-indigenous aquatic organisms (such as zebra mussels, Eurasian milfoil, river ruffe, and other species) are closely coordinated with appropriate state, federal, and tribal management authorities.

Habitat Restoration and Improvement

The WDNR is responsible to the People of the State of Wisconsin for protecting and maintaining all habitats capable of supporting aquatic species. The department administers programmes for deterring point and non-point pollution, controlling vegetation and rough fish, regulating and controlling lake and reservoir water levels and limiting shoreline development. Where economically and ecologically feasible, the department improves fish habitat by developing instream and riparian habitat improvement structures such as wing deflectors, bank riprap, stream bank fencing and fish shelters. Dredging and streamside brushing may also be practiced for the same purpose. Where relevant, the application of these techniques must be consis-tent with the wild and wilderness policies of the State.

Wild and wilderness lakes and streams are a special and limited resource providing unique settings for enjoyment of fishing and other outdoor activities. Special management methods that increase fishing quality are encouraged on these waters. Such methods may include trophy fishing, regulated harvest, special seasons and controlled entry.

NATIONAL PARK SERVICE

Fisheries management in the National Park System is directed by Servicewide policy and guidelines that find roots in the NPS's founding legislation, the NPS Organic Act of 1916 (16 USC 1). Many individual parks also have specific fishery stipulations within their enabling legislation that further direct fisheries management within these units. The Organic Act directs the Secretary of the Interior and the NPS to manage national parks and monuments to "conserve the scenery and the natural and historic objects and the wild life therein and to provide for the enjoyment of the same in such manner and by such means as will leave them unimpaired for the enjoyment of future generations."

These general powers were broadened by the Redwood Expansion Act of 1978 (16 USC 1a1) in which Congress gave further direction to the Secretary to ensure that the management and administration of the National Park System "shall not be exercised in derogation of the values and purposes for which these various areas have been established, except as may have been or shall be directly and specifically provided by Congress."

Consistent with these broad authorities, the current NPS guideline for fisheries management emphasizes the restoration and preservation of natural assemblages of native species within natural areas. The NPS manages all park resources with an emphasis on fundamental ecological processes as well as for individual species and communities. Fisheries management programmes strive to preserve or restore the natural behaviour, genetic variability and diversity, and ecological integrity of fish populations. The NPS's recreational fisheries programme, "A Heritage of Fishing," provides a framework for improving recreational fishing and management of fishery resources while incorporating the.NPS's fundamental mandate to preserve and restore these fisheries. NPS fisheries management programmes are consistent with the 1988 National Recreational Fisheries Policy, Executive Order 12692 on Recreational Fishing, and the 1996 Interagency Recreational Fishery Resources Conservation Plan. Fisheries and aquatic resources of the Riverway have been and continue to be influenced by human activities within the watershed.

The Riverway comprises a landmass of approximately 97,850 acres within a watershed area of approximately 4,950,000 acres. NPS authority to manage the watershed, water quality, and fisheries is limited to a narrow ribbon of land along the 252 miles of the St. Croix and Namekagon Rivers. Virtually all of the tributaries to these rivers within the Riverway have origins or significant drainages beyond the Riverway's boundary.

Consequently, cooperation and collaboration with state and local governments and private landowners will be critical to the longterm sustainability of Riverway aquatic resources.

Recreational Fishing

Section 13(a) of Public Law 90542 establishing the St. Croix National Scenic Riverway, specifically permits hunting and fishing on lands and waters under the Riverway's jurisdiction in accordance with applicable Federal and State Laws. NPS Management Policies allow for recreational fishing in parks where it is authorized by federal law or where it is not specifically prohibited, and does not interfere with the functions of natural aquatic ecosystems or riparian zones. In addition, these policies require that any restrictions on recreational uses will be limited to the minimum necessary to protect park resources and to promote visitor safety and enjoyment.

However, recreational fishing may be restricted by the NPS at any time, after consultation with the States, to achieve park resource management objectives, to enhance or facilitate public safety and administration or to accommodate public use and enjoyment. The Organic Act also grants the Secretary the authority to implement "rules and regulations as he may deem necessary or proper for the use and management of the parks, monuments and reservations under jurisdiction of the NPS (16 USC 3)." Fishing regulations in 36 CFR Part 2.3 apply on lands and waters that are within park boundaries and under the legislative jurisdiction of the United States, regardless of ownership. In addition, in parks where NPS has proprietary jurisdiction, such as at St. Croix

NSR, state laws and regulations also apply to the fishery. However, NPS retains authority to implement more restrictive regulations (16 USC 3).

Due to the complexity of the regulatory framework, close collaboration among the parties will be necessary to provide Riverway visitors with consistent sportfishing and recreational use regulations. Differences in regulations governing minimum fish size, seasons, creel limits, and manner of taking should be documented and assessed. Such assessments will guide managing agencies in the cooperative development of consistent regulations. Sportfishing and recreational use regulations need to be evaluated on a periodic basis for not only interjurisdictional consistency, but to assess their effectiveness as a resource management tool. Such assessments need to be closely correlated with fish population monitoring and creel census data. Regulations may provide for reasonable use and enjoyment of the resource but should not compromise the productivity and sustainability of the fisheries or their resource base. Fishery management objectives should be based on maintaining biological balance and integrity as well as the quality of the fishing experience.

Fish Stocking

Fish stocking has been a traditional management tool of the NPS since the early days of fisheries management in Yellowstone National Park in the late 1800's. Present day fish stocking policies were developed to achieve the NPS's broad resource management objectives. These policies place significant constraints on the use of stocking as a management tool. Fish stockings are generally approved for use in fisheries restoration projects or to restore and/ or enhance recreational fisheries in certain park management zones. Most parks of the National Park System are classified in management categories recognizing that various types of parks have different purposes requiring management strategies specifically tailored to address their management mandates. Also, individual parks may be zoned into subunits requiring different management strategies. The four primary management categories/zones include natural (including wilderness), cultural, park development, and special use. Fish stocking policies are most restrictive in natural and most liberal in special use zones. Special uses zones often include altered aquatic environments such as reservoirs, humanmade ponds, and altered streams. In areas where exotic species have become well established, self-sustaining and support important recreational fisheries, the superintendent may request a waiver of policy from the Director of the NPS to continue supplemental stockings.

Exotic and Native Fishes

An exotic species, by NPS definition, is any species whose presence is the result of direct or indirect, deliberate or accidental, actions by humans (NPS Management Policies 4:11, 1988). Any species establishing selfsustaining populations outside of its natural range is considered an exotic. Species such as brown trout (*Salmo trutta*), carp (*Cyprinus carpio*), and zebra mussels (*Dreissena polymorpha*) are clearly exotics by this definition since none of these are native to North America. The NPS does not recognize "naturalized" species. Rainbow trout (*Oncorhynchus mykiss*), which are native to western North America, would also be considered an exotic species in the Riverway by NPS definition.

Determination of the need for control or eradication of an exotic species requires consideration of type and severity of threats to native species, public health, and whether or not control measures would be prudent and feasible. In instances where control measures would be impractical or where wellestablished exotic fish species are ecologically balanced within the aquatic community, highly desirable to the recreational angler and strongly supported by the public, any actions that encourage survival must be permitted under a waiver of policy by the Director of the NPS and must be included in the park's Resource Management Plan or Fishery Management Plan.

Any decision to restore native fish populations at the expense of exotic fish must include an evaluation of several factors including: 1) the presence of or potential for restoring habitat, 2) the availability of the historic genotype, 3) the degree of humaninduced habitat alterations, 4) the feasibility of the effort, and 5) a requirement that the restored native species be selfsustaining. In general, efforts to control or eradicate selfsustaining exotic fish populations are costly and ineffective, particularly in open aquatic systems such as rivers and streams where fish are free to immigrate and emigrate. Exceptional care should be taken to understand and predict community and ecosystem level changes resulting from any efforts to control or eradicate established exotic fish populations.

Habitat Restoration and Improvement

Habitat manipulations or any interference with natural ecological processes in park natural zones are allowed only when directed by Congress, in emergencies involving life or property or to restore ecosystem functions. In special use zones, habitat manipulations may be carried out to restore or enhance fish populations for the purpose of maintaining or restoring recreational fishing. These habitat manipulations may include, but are not limited to, stream improvement structures, fencing, and riparian zone restoration.

For the Riverway, the use of habitat restoration techniques, along with longterm interagency watershed protection programmes, may be required to restore native brook trout populations in the coldwater zone. Improvements in riparian habitat and overall watershed conditions, as well as fishery oriented management of oldaged forest stands, will be necessary to reduce ambient temperature and brown trout populations and restore brook trout. Because of requirements of the Wild and Scenic Rivers Act, habitat restoration projects should not alter streambeds and banks in such manner as to measurably change the natural freeflowing characteristics of the river. When practical and ecologically and economically feasible, humanmade obstructions to the rivers' freeflowing character, such as lowhead dams and dikes, should be considered for removal and reclamation.

POLICY CHANGE OF PARK MANAGEMENT

The major policy change which we would recommend to the National Park Service is that it recognize the enormous complexity of ecologic communities and the diversity of management procedures required to preserve them. The traditional, simple formula of protection may be exactly what is needed to maintain such climax associations as arctic-alpine heath, the rain forests of Olympic peninsula, or the Joshua trees and saguaros of southwestern deserts. On the other hand, grasslands, savannas, aspen, and other successional shrub and tree associations may call for very different treatment. Reluctance to undertake biotic management can never lead to a realistic presentation of primitive America, much of which supported successional communities that were maintained by fires, floods, hurricanes, and other natural forces.

A second statement of policy that we would reiterate -- and this one conforms with present Park Service standards -- is that management be limited to native plants and animals. Exotics have intruded into nearly all of the parks but they need not be encouraged, even those that have interest or ecologic values of their own. Restoration of antelope in Jackson Hole, for example, should be done by managing native forage plants, not by planting crested wheat grass or plots of irrigated alfalfa. Gambel quail in a desert wash should be observed in the shade of a mesquite, not a tamarisk. A visitor who climbs a volcano in Hawaii ought to see mamane trees and silver-swords, not goats.

Carrying this point further, observable artificiality in any form must be minimized and obscured in every possible way. Wildlife should not be displayed in fenced enclosures; this is the function of a zoo, not a national park. In the same category is artificial feeding of wildlife. Fed bears become bums, and dangerous, Fed elk deplete natural ranges. Forage relationships in wild animals should be natural.

Management may at times call for the use of the tractor, chain-saw, rifle, or flamethrower but the signs and sounds of such activity should be hidden from visitors insofar as possible. In this regard, perhaps the most dangerous tool of all is the roadgrader. Although the American public demands automotive access to the parks, road systems must be rigidly prescribed as to extent and design. Roadless wilderness areas should be permanently zoned. The goal, we repeat, is to maintain or create the mood of wild America. We are speaking here of restoring wildlife to enhance this mood, but the whole effect can be lost if the parks are overdeveloped for motorized travel. If too many tourists crowd the roadways, then we should ration the tourists rather than expand the roadways.

Additionally in this connection, it seems incongruous that there should exist in the national parks mass recreation facilities such as golf courses, ski lifts, motorboat marinas, and other extraneous developments which completely contradict the management goal. We urge the National Park Service to reverse its policy of permitting these non-conforming uses, and to liquidate them as expeditiously as possible (painful as this will be to concessionaires).

Another major policy matter concerns the research which must form the basis for all management programmes. The agency best fitted to study park management problems is the National Park Service itself. Much help and guidance can be obtained from ecologic research conducted by other agencies, but the objectives of park management are so different from those of state fish and game departments, the Forest Service, etc., as to demand highly skilled studies of a very specialized nature.

Management without knowledge would be a dangerous policy indeed. Most of the research now conducted by the National Park Service is oriented largely to interpretive functions rather than to management. We urge the expansion of the research activity in the Service to prepare for future management and restoration programmes. As models of the type of investigation that should be greatly accelerated we cite some of the recent studies of elk in Yellowstone and of bighorn sheep in Death Valley. Additionally, however, there are needed equally critical appraisals of ecologic relationships in various plant associations and of many lesser organisms such as azaleas, lupines, chipmunks, towhees, and other non-economic species.

In consonance with the above policy statements, it follows logically that every phase of management itself be under the full jurisdiction of biologically trained personnel of the Park Service. This applies not only to habitat manipulation but to all facets of regulating animal populations. Reducing the numbers of elk in Yellowstone or of goats on Haleakala Crater is part of an overall scheme to preserve or restore a natural biotic scene. The purpose is single-minded. We cannot endorse the view that responsibility for removing excess game animals be shared with state fish and game departments whose primary interest would be to capitalize on the recreational value of the public hunting that could thus be supplied. Such a proposal imputes a multiple use concept of park management which was never intended, which is not legally permitted, nor for which can we find any impelling justification today.

Purely from the standpoint of how best to achieve the goal of park management, as here defined, unilateral administration directed to a single objective is obviously superior to divided responsibility in which secondary goals, such as recreational hunting, are introduced. Additionally, uncontrolled public hunting might well operate in opposition to the goal, by removing roadside animals and frightening the survivors, to the end that public viewing of wildlife would be materially impaired. In one national park, namely Grand Teton, public hunting was specified by Congress as the method to be used in controlling elk. Extended trial suggests this to be an awkward administrative tool at best.

Since this whole matter is of particular current interest it will be elaborated in a subsequent section on methods.

FISHERIES MANAGEMENT PLANNING

Fisheries management planning involves the development of a plan which describes a particular fishery and its particular problems, the actors involved in the fishery and the objectives for the development of the fishery, and which outlines the measures for control of fishing to ensure the sustainable utilization of the fishery resource. The management plan may be linked to mechanisms that limit entry or restrict fishing. For example, some of the provisions might provide that a licence may be refused on the grounds that to issue it would undermine the objectives of a fishery plan. The fisheries management plan should be developed with input of technical experts, and provisions should be included to monitor its implementation. For many countries, the development of a fishery plan used to be an administrative process and the prerogative of the fisheries management authority. In such countries, fishery management planning was a good management technique but not a mandatory requirement. In recent years, however, more countries have embraced the management planning concept and have eventually required the development of fishery plans in legislation. The Australian state of Queensland prescribed the development and implementation of management plans for the state's major fisheries in legislation in 1994.

Papua New Guinea first introduced the use of fishery plans in the same year. Nauru introduced its version of fishery plans, the "fisheries strategy", in 1997. Solomon Islands legislatively introduced fishery management and development plans for the first time in 1998. Malawi and Senegal also legislatively introduced fisheries management plans in legislation in 1997 and 1998, respectively. These are but a few of the many countries that have adopted the use of fishery plans in fisheries management in the last decade.

INCREASED PARTICIPATION AND DEVOLUTION OF FUNCTIONS

National fisheries management, like many natural resource management regimes, heavily concentrated management authority in the central (national) government or its agencies. In recent years, however, the effectiveness of these top-down management approaches has been questioned.

The main criticisms include:

- The lack of consultation with stakeholders or with the regulated, which results in the lack of a legitimate basis for regulations or management measures and, consequently, non-compliance;

- Implementation and enforcement in such systems relies heavily on adequate technical capacity and other resources, of which there are never enough, thus adding to the high rate of non-compliance;
- Resource users do not appreciate that the sustainable use of the fisheries resources is vital to their livelihood or the nation's economy because they do not feel part of or "owners of" the management process and measures.

Recent legislation has expanded the scope of involvement of stakeholders so that there is broader participation both at the decision making and implementation levels. The expansion of involvement of stakeholders has been in three main areas:

- Consultation, whereby the management authorities solicit the views of persons who are interested in or could be affected by the management decision, so that their views can confirm or cause an amendment to the proposed management decision or regulation as appropriate;
- Formal representation of stakeholders on consultative, advisory or decisionmaking institutions within the fisheries management framework;
- Devolution of management or implementation powers, or both, to lower-level governments and stakeholder communities or groups. Examples of the expansion of participation in the consultation process can be found in the fisheries legislation of Barbados, New Zealand and the Philippines. In the context of fisheries management plans, it is argued that the participation of stakeholders in the preparation of such plans increases the opportunity for their effective implementation.

Broad participation through formally established institutions is a more common practice than participation through direct consultation or devolution of powers. Barbados and Mauritius facilitate such participation through Advisory Committees, whereas Malawi has an Advisory Board and South Africa has Consultative Advisory Forums. In formerly communist regimes, Albania established Central and Local Consultative Commissions whereas Lithuania has a Fisheries Board.

Participation of stakeholders by means of the third approach (*i.e.*, devolution of powers) is a more progressive form of participation in fisheries management compared to the other two. It is also the most difficult to institutionalize because it involves laws relating to political governance which may implicate the constitution or other fundamental laws on governance or decentralization.

Nevertheless, recognition of the important role to be played by communities whose livelihoods depend on fishing, and transfer of fishery management functions to such communities or to lower-level governments, has been slowly gaining popularity (as well as controversy) over the last decade. Canada, New Zealand and the United States, which have been negotiating with native or aboriginal communities for years, have developed schemes to recognize or give deference to native rights to fish or manage fisheries. In Japan, fisheries cooperatives have enjoyed some form of exclusivity in fishing and management over coastal resources falling within their domain.

The Philippines has been a leader in the developing world in the devolution of fisheries management powers. The government formalized the decentralization of fisheries management powers to municipalities in 1991 through legislation and consolidated it in subsequent legislation in 1998. In the Marshall Islands, Local Government Councils are responsible for the management, development and sustainable use of the reef and in-shore fisheries, extending up to five miles seaward from the baseline from which the territorial sea is measured.

A variant of devolution regimes is co-management, as seen in Malawi. There, since 1993, participatory fisheries management programmes have been introduced whereby local-level institutions and the Department of Fisheries jointly make decisions. This scheme has replaced the former centralized management system.

FOOD SAFETY REGULATION FOR FISH AND FISH PRODUCTS

In many developing countries, seafood exports are an important source of revenue and there is growing interest in government policies and practice relating to fish and fish product exports. Although food safety is generally the purview of ministries or agencies concerned with food and human health, in some countries government may decide that seafood safety is best regulated by the authorities responsible for the fisheries sector, and this may require legislative action.

Food safety issues affect fisheries particularly in the post-harvest sector. The main trend is the introduction of the Hazard Analysis and Critical Control Point (HACCP) system into the European Union (EU) and the United States, and its recognition as a food safety assurance mechanism by Codex Alimentarius. The EU and the United States have both made fish and fishery products the first category of foods in the food industry subject to mandatory application of HACCP systems. The EU issued the first regulation for fish products "laying down the health conditions for the production and the placing on the market of fishery products" in 1991. In May 1994, the EU adopted an additional regulation which made it mandatory to impose more precise rules for the application of health checks. The United States adopted a seafood HACCP regulation, Procedures for the Safe and Sanitary Processing and Importing of Fish and Fishery Products, in December 1997.

These developments have caused the major seafood-exporting countries to adopt or meet the standards established by the importing countries. For example, Namibia and South Africa, which have significant exports to the EU, require that seafood processors and exporters meet Directive 91/493. The directive concerns both domestic (EU) and third country (non-EU) production. It defines EU standards for handling, processing, storing and transporting fish. It must be noted that processed bivalve molluscs (as well as tunicates, marine gastropods and echinoderms) are subject to both Directive 91/492 and Directive 91/493.

Directive 91/493 lays down rules on conditions applicable to factory vessels, on-shore plants, packaging, storage and transport. Provisions that may require more details are set concerning auto-controls, parasites (all visible parasites must be removed), organoleptic, chemical and microbiological checks. National authorities responsible for standards, food safety and fisheries management are also charged with encouraging exporters to meet HACCP and ISO 9000 standards and guidelines, with the ultimate objective of formally adopting these systems through regulation.

In Tonga, seafood exporters to the United States have taken it upon themselves to implement HACCP and other applicable requirements of the seafood safety regulations administered by the U.S. Food and Drug Administration. It is a matter of Tonga government policy that regulations for the processing of seafood and a certification system for seafood exports are to be introduced soon. To this end, draft seafood safety regulations are currently under consideration for promulgation under the principal fisheries legislation. The United Republic of Tanzania is one of the few countries to adopt regulations (the Fish Quality Control and Standards Regulations of 2000) to specifically implement HACCP.

LEGISLATIVE IMPLEMENTATION OF INTERNATIONAL FISHERIES INSTRUMENTS

The 1993 FAO Compliance Agreement and the 1995 UN Fish Stocks Agreement address the nature of state obligations at the international level; both agreements seek to define with a degree of specificity which is unusual in global fisheries agreements the precise ways in which state parties should meet those obligations. Implementing legislation will be necessary in most countries to meet these requirements as a precondition to state ratification or acceptance. This will usually call for new legislation, although, depending on the legislation already in place, some countries may be able to implement the Agreements through changes to current legislation or through subsidiary legislation.

Compliance Agreement

The Compliance Agreement reinforces the effectiveness of international fisheries conservation and management measures by redefining and reinforcing the concept of flag state responsibility for the activities of fishing vessels flying the flag of a state party. It also seeks to provide means to ensure the free flow of information on all high seas fishing operations.

The agreement requires the following issues to be implemented in the national legislation of its parties:

- Designation of the national authority responsible for carrying out the duties of the flag state under the agreement;
- Provisions that make it unlawful for flag vessels to undermine the effectiveness of international conservation and management measures, and that provide a mechanism for authorities to ensure that the law is respected;
- Mandatory fishing authorizations for flag vessels fishing on the high seas;
- Mandatory conditions for flag vessels receiving a fishing authorization;
- Proper marking of fishing vessels;
- Information on fishing operations;
- Enforcement measures and sanctions;
- Establishment and maintenance of records of flag vessels fishing on the high seas;
- Duties of the flag state to provide FAO with information.

The Compliance Agreement has not yet entered into force but the parties to the agreement have started to give effect to it through legislation. At the time of writing, the following countries had enacted legislation or made amendments to existing legislation to implement the provisions of the agreement: Australia, Canada, Namibia, New Zealand, Norway, Seychelles, South Africa, Saint Vincent and the Grenadines and the United States. The style with which these countries have legislated on the essential components of the Compliance Agreement, and the scope of their legislative action, varies as the countries attempt to reflect their national situations and respond to their particular needs. Time does not permit an examination of the details of these national efforts, although it can be said that these countries have clearly charted the general approach on how to incorporate the requirements of the agreement into national legislation.

UN Fish Stocks Agreement

The UN Fish Stocks Agreement entered into force in December 2001.

The main elements that need legislative implementation are:

- Coastal states and distant water fishing states are required to ensure that the conservation and management measures which are created within the exclusive economic zone (EEZ) and on the high seas are compatible.

- Parties to the agreement are to apply general principles for the conservation and management of straddling fish stocks and highly migratory fish stocks, including the precautionary approach, on the high seas as well as within their EEZ.
- Flag states must meet certain obligations with respect to their vessels fishing on the high seas for straddling fish stocks and highly migratory fish stocks.
- State parties are obliged to join regional fisheries management organizations or arrangements, or to agree to comply with the conservation and management measures those bodies create. Otherwise they will not be allowed to fish in the areas where such measures apply.
- Non-flag states are subject to innovative enforcement provisions, and a new concept of port-state jurisdiction in respect of fishing vessels applies.
- States will have to apply detailed provisions on peaceful dispute settlement.

The agreement is complex and therefore difficult for states to legislate on. Nevertheless, a number of countries have enacted legislation to implement it, namely: Australia, Canada, Iceland, Namibia, New Zealand, Norway, Seychelles, South Africa, Saint Vincent and the Grenadines and the United States. In most cases, both the Compliance Agreement and the UN Fish Stocks Agreement are implemented through the same legislation, as in New Zealand.

ECOSYSTEM-BASED MANAGEMENT

In the recent years, the fishery science started internalizing the notion that management by single species and only by controlling fishing is illogical and contrary to ecological realities. A new approach entered the domain of fisheries management under the name of "Ecosystem-based management". So what the new approach should be about? Let's consider the main elements of an ecosystem affected by commercial fishery (fishery ecosystem) and their changes in space, time and character.

The first element is the *time factor*. Any aquatic ecosystem is a dynamic, pulsating, and ever-changing macro-organism. Thus, trends and fluctuations that have been occurring in the ecosystem throughout history must be taken into account. The second element embraces all the changes and physical, biological and chemical forces, including upstream and coastal pollution, imposed upon the ecosystem by *climatic variations* and the various *human activities*. Whatever happens with inshore and bottom habitats, and the water, affects the ecosystem's biota (*i.e.*, all living things). The third element is made up of the *relationships* between the various species occupying the ecosystem at all stages of their life. This runs from bacteria and phyto-plankton up to top predators, with special attention to the managed "target species", physical factors affecting them, their food, prey and predators. Last, but not least is *fishing*, which apart from massive removal of marine organisms from the system, also may influence the size, age and genetic composition of the fished populations. The fishing itself is influenced by the market and the socio-economic context of fishing people and their communities, and of fish consumers, as well as by the industries and technologies involved. No doubt, *controlling fish harvests is not enough to ensure sustainable fishery and healthy ecosystems*.

"The vastness of linkages between species and critical habitats in a coastal area requires comprehensive management of all its parts". An appropriate approach should incorporate institutional arrangements and cultural factors to provide for better analysis and prediction. The list of the factors other than fishing that affect the size, composition and well-being of populations of marine organisms would be extensive. Here're some examples:

Further expansion of dead anoxic zones, 150 of which were reported in 2003 in bays and semi-enclosed seas, some of which extended up to 27,000 square miles, may well be a greater peril than overfishing. A slowly killing mycobacteriosis epidemic was affecting the condition and size of the striped bass population in Maryland and Virginia and in the heavily polluted Chesapeake Bay. According to some estimates, the word's seabirds consume 70 million mt of food as compared to the 80-90 million mt of global marine landings, and the amount of fish eaten by marine mammals worldwide is several times the worldwide ocean fish harvest.

Climatic fluctuations and separate events cause boom-and-bust sequences among fish populations and other marine organisms. There may be umpteen sorts of association between the physical and biological elements. For example, studies show that the main factors critical for salmon abundance are distance to and size of oceanic areas with temperature and food critical to the survival of juveniles, such as the abundance of krill, which is their main food. Pollution of rivers and estuaries and longshore development that destroys inshore habitats, affect the reproduction of many fish species whose spawning and nursery areas are in inshore waters.

FISHERIES SUSTAINABILITY

In most coastal areas, rehabilitation of marine habitats and restoration or protection of essential upstream environments and coastal hydrology functions are essential to future fisheries sustainability. The fundamental message is that for such sustainability we must restructure general coastal management. Environmentalist NGOs and everyone involved in trying to secure sustainability of habitats must realize that there has to be a balance between the needs of the people and the conservation of living resources, and both sides need to be prepared to compromise. The environmentalists' and some scientists' advice that by only controlling fishing they'd achieve sustainability is counter-productive when it diverts attention from other, often more significant factors.

Only an intelligent *analysis and synthesis* of the interaction of all important elements involved can enable understanding of the *problematique* of ecosystem management and searching for solutions. We have to accept that the dynamics of the system is so complex that even the best existing models are of limited use for forecasting the outcomes of actual management actions. In this general context, "*ecosystem-based fishery management*" must address the particular elements. Those that require intervention must be defined and managed respectively, while the others, including those that are beyond the management's powers, must be taken into account. Flexibility and tailoring management for each specific fishery and area should become the rule. There's no rational way to impose a policy that doesn't take all the above into account, or a single "common" policy for different fisheries, ecosystems and fishing cultures. We have to analyze separately each fishery ecosystem to see, for example, if TACs are the right methodology to use or other options such as effort control should be considered.

Unfortunately, some authors and "green" zealots are trying to hijack the concept of "ecosystem management" by representing it as care-taking of the fishery that it doesn't affect the ecosystem. While paying lip service to other factors, they focus on commercial fishing in a way which stigmatizes it as the only factor affecting ecosystem that needs tackling and as the main villain to be constrained or eliminated in order to "protect" marine ecosystems. Their interest doesn't lie in cleaning up habitats and keep fisheries prosperous, but elsewhere. Where? – This depends on who's the "father" (or sponsor) of the particular "stewardship". The institutional and human inertia is strong and many NGOs as well as the fishery management system worldwide abound in scientists and environmentalists who preach sustainability, but keep talking of "saving the oceans" just by shutting down fisheries or by closing off major sea areas to fisheries, and hardly even mention the whole set of other vectors, whatever may be their relative roles in each separate ecosystem. They still ignore physical and biological-ecological environmental factors and upstream and marine pollution from industrial, municipal, and agricultural sources. They close the eye on inshore habitat destruction, and disregard or play down non-human predation, nowadays enhanced by the extensive protection bestowed upon marine mammals.

In fact, however, ecosystem management approach places fishing in its proper context, so it can't be blamed anymore for everything that happens in the sea. Ecosystem management doesn't allow managers and politicians to keep turning a blind eye to pollution, environmental variations, habitat destruction, etc. It carries fishery management from the current over-simplification into more complex concepts. This obviously worries some scientists and managers, who are concerned that with "ecosystem management" they'd be sailing in unknown waters and lose their control over fishing rights and quotas.

Doubtless, it won't be easy, but the point is that to manage fishery ecosystems using the existing methodologies, as they've practically been since the mid 20th century, is like looking under a streetlamp for a lost key, because that's where the light is, instead of looking for it where it may be-however dark is the place. The new approach may also require to get used to the fact that for a long time to come we won't be able to put numerical values to all acting factors, a current custom that sometimes borders on numerology, and relearn to talk also in qualitative terms and perhaps to employ fuzzy logic.

While ecosystem-based management is coming of age, institutional inertia and obstinacy, human conservatism, fear of the unknown and need to admit own misconstruction of the realities of the coastal, marine and living resources, impedes its application. The notion of positive development stems from the recognition that, climatic conditions permitting, sustainability of both, marine capture fisheries and marine aquaculture, can only be achieved by integrated and balanced approach to ecosystem management involving upstream and downstream influences, fisheries, and any other human activities that affect coastal and offshore habitats. There is a growing understanding of the complexity of the systems to be managed and of the futility of simplistic approaches, such as those of the still prevailing fishery management based on inadequate science and its mathematical models.

THE WORKING PRINCIPLES OF FISHERIES MANAGEMENT

A complex and possibly confusing picture of all the tasks which need to be considered by the fisheries manager. Some of this complexity can be reduced by attempting to highlight the underlying key issues. There are both benefits and risks in attempting to simplify a subject and over-simplification can lead to neglect of important details. However, simplification can facilitate understanding important principles and highlighting the broad areas which need attention. Arising from the considerations, a number of key principles can be identified which may serve to focus attention on the starting points for effective fisheries management.

WHO IS THE FISHERY MANAGER

The Technical Guidelines (FAO, 1997) suggest that fisheries management institutions have two major components: the fisheries management authority and the interested parties. The fishers and fishing companies would usually be the major participants amongst the interested parties. The fisheries management authority is that entity which has been given the mandate by the

State (or States in the case of an international authority) to perform specific management functions.

In many countries that authority would be a Department of Fisheries or, within a broader Department, a Division of Fisheries. However, a fisheries management authority does not have to fall directly within central government, and could be, for example, provincial, local, parastatal or private. Any one of these arrangements can function effectively, given an adequate legal framework in which to operate and the resources necessary to fulfil their function.

Who, then, within this authority is the fisheries manager and to whom is this Guidebook addressed? In fact, despite the fact that we have deliberately used the term in the title, we suggest that in modern fisheries management, there is rarely a single individual who fulfils the functions of "fisheries manager". The head of the authority, for example, a Director of Fisheries, may have overall responsibility for implementing fisheries management and, as well as being accountable and responsible for the advice passed on from his or her Department to the political decision-maker, may act in an overall coordinating role.

However, this individual is unlikely to, and generally should not, have sole responsibility for receiving information, formulating advice and making and implementing decisions. Fisheries management is a complex and multi-faceted discipline and requires input from a range of perspectives. It is therefore inappropriate to expect any individual to fulfil this function on their own. Fisheries management should involve the legitimate interested parties in the management process.

Perhaps the closest we can come to a "fisheries manager" is the management authority as a whole, including technical experts, monitoring, control and surveillance (MCS) units, administrative units, the executive body of the formal authority, the consultative mechanisms, the advisory body where one exists, and the responsible political head who is often a Minister. Each member of these functional bodies is, to some extent, a fishery manager and this Guidebook is aimed at all of them. It is not designed to go into great technical and operational detail on each function or task as this would require a set of Guidebooks. Instead, it is intended to give a holistic picture of how the different functions should interact in a fisheries management authority in order to develop appropriate objectives, management strategies and plans, and how to encourage all those participating in a fishery to collaborate in and adhere to the agreed strategy.

CONSTITUTES A MANAGEMENT AUTHORITY

The responsibility for fisheries management rests with the designated fisheries arrangement or organisation which, in this Guidebook, we have referred to without distinction as the fisheries management authority. Following the practice used in the Technical Guidelines on Fisheries Management (FAO,

1997) the term is used broadly here to describe that legal entity which has been designated by the State as having the mandate to perform specified fisheries management functions. In practice, it may be a national or provincial ministry, a department within a ministry, or an agency and could be governmental, parastatal or private. In the case of shared resources it should be international.

The area of competence, geographical area, fish resources and fisheries for which a given management authority is responsible must be precisely specified in each case in the appropriate legislation. The task of an authority is diverse and complex and as a result, fisheries management authorities are normally divided into institutional support structures: the fisheries management institutions. The institutions need to encompass the basic tasks and functions of fisheries management. The actual institutional structure and mechanisms may differ from authority to authority and it would be inappropriate for us in this Guidebook to attempt to prescribe any specific set of characteristics as representing the 'best' institutional structure and processes. What is best in each case will depend in large part on the specific circumstances and context. What is universal, however, is that it is essential for the different institutions concerned with management of any fishery or fisheries to be able to interact effectively, requiring good channels of communication and feedback. The institutions must also be seen by the different interested parties as being legitimate. The need for collaboration between the authority and the interested parties is as important as collaboration between the institutions within the authority. Examines the pre-requisites for effective partnerships between the management authority and the interested parties and the different types of partnership which can be considered.

It is common and frequently desirable for the national government to devolve all or some fisheries management functions to local government or to smaller groups such as fishing communities. In such cases, it is essential to specify precisely the responsibilities and functions, including the geographical area, falling under this local authority or smaller group. The institutions within the local authority must follow the same principles as those for a national authority.

The Code of Conduct requires that fisheries management should be concerned with the whole stock over its entire area of distribution and therefore that States should cooperate in the management of transboundary, straddling, highly migratory and high seas fish stocks exploited by two or more states.

General rules for cooperation towards conservation of such fish stocks are foreseen in the United Nations Convention on the Law of the Sea of 10 December 1982, and in the 1995 UN Fish Stocks Agreement. The responsibilities, functions and structure of international or regional fisheries authorities will usually not differ substantively from those of national authorities.

GOALS AND OBJECTIVES

The over-riding goal of fisheries management is the long-term sustainable use of the fisheries resources. Achieving this requires a proactive approach and should involve actively seeking ways to optimise the benefits derived from the resources available.

This rarely happens, though, and fisheries management is still most commonly practised as a reactive activity, where decisions are made and actions taken largely in response to problems or crises. The resulting crisis decisions are then normally attempts merely to solve the immediate problems without properly considering the broader perspective and the longer-term objectives. Such an approach may succeed in maintaining dissatisfaction sufficiently low to avoid major conflict, but it is extremely unlikely to result in the best use of the marine resources being exploited by the fishery.

The first step in proactive fisheries management is to decide what is meant by optimising the benefits for each fishery - what can the State or the collection of legitimate interested parties agree on as being optimal benefits? This may be described in general terms in the national fisheries policy which must be the starting point for determining the specific objectives for each fishery. The broad goals stated in the fisheries policy may need to be tailored for a specific fishery, but the goals for each fishery should be consistent with the policy.

In general terms, the goals in fisheries management can be divided into four subsets: biological; ecological; economic and social, where social includes political and cultural goals. The biological and ecological goals may be more correctly thought of as constraints in achieving desired economic and social benefits but for simplicity and consistency with the terminology most commonly used in fisheries management, we will include them as goals in this Guidebook. Examples of goals under each of these categories include:

- To maintain the target species at or above the levels necessary to ensure their continued productivity (biological);
- To minimize the impacts of fishing on the physical environment and on non-target (bycatch), associated and dependent species (ecological);
- To maximize the net incomes of the participating fishers (economic); and
- To maximize employment opportunities for those dependent on the fishery for their livelihoods (social).

Identifying such goals is important in clarifying how the fish resources are to be used to benefit society, and they should be agreed upon and recorded, both at the policy level and for each fishery. Without such goals, there is no guidance on how the fishery should be operated, which results in a high probability of *ad hoc* decisions and sub-optimal use of the resources (resulting in lost benefits), and increases the probability of serious conflicts as different interest groups jostle for greater shares of the benefits. This is often seen in practice and one of the important causes of failures in fisheries management has been identified as the frequent absence of clear and precise objectives.

While setting goals is an essential first step, the goals stated above have two obvious limitations. Firstly, they have clear conflicts in intention as it is impossible, for example, to minimize impacts of the fishery on the ecosystem and simultaneously to maximize net incomes. Similarly, it is very probable that management strategies that aim to maximize net incomes will not also maximize employment opportunities.

Some compromise between these goals has to be achieved before an effective management strategy can be devised. The second limitation of the goals is that they are too vague to be of much benefit to the manager. For example, the impacts of fishing can only be "minimized" by having no fishing at all, which is unlikely to have been the intention of those who stated the goal. Maximising employment opportunities could mean allowing as many fishers as possible to participate, regardless of whether or not they could make a living from the fishery, or it could mean maximising the number which could still earn some acceptable income, or many other such targets. Too much is left to the discretion of the manager with these examples of goals.

It is therefore necessary to refine the goals further and to develop operational objectives for each fishery. Operational objectives are very precise and are formulated in such a way that they should be simultaneously achievable in that fishery. In other words, the trade-offs between the biological, ecological, economic and social goals must have been agreed upon and the conflicts and contradictions resolved. The development of operational objectives is illustrate the difference between goals and operational objectives, two examples of objectives are:

- To maintain the stock at all times above 50 per cent of its mean unexploited level (biological);
- To maintain all non-target, associated and dependent species above 50 per cent of their mean biomass levels in the absence of fishing activities (ecological).

With operational objectives such as these, it is possible for any observer, including the manager, to establish whether or not they are being achieved and hence whether or not the management strategy is appropriate and being successfully implemented. These operational objectives can also easily be used as the foundation for reference points, which are essentially the operational objectives expressed in a way which can be estimated or simulated in a fisheries assessment. Once operational objectives have been agreed upon, a management strategy can be developed, made up of a suite of different management measures, to achieve those objectives.

All of this may sound complex, but in reality is no more than most people do in order to develop a budget for their personal finances. Most of us have realistic but imprecisely expressed hopes and needs for our lifestyle as well as a knowledge of the nature of the resource (in this case our net income). These hopes and needs are the goals of our budget but they will all compete for the same resource, our net income, so there will probably be conflicts which need to be resolved. Therefore we have to modify our goals and express them more precisely: we develop operational objectives in which we specify what we can realistically achieve in terms of food, housing, education etc. Thereafter, we need to decide on our budget strategy: how can we meet those objectives: what type and quantities of food and clothing should we be buying; what type of housing can we consider; can we consider an annual holiday, etc.

Clearly, our operational objectives must be consistent with the yield we can expect from the resource (our income). Normally the process of developing realistic objectives will require trade-offs and most of us find, for example, that we cannot allocate as much for entertainment or holidays as we would like and at the same time make our rental or mortgage payments. Therefore priorities are established and compromises made until eventually we arrive at realistic objectives that balance our desires with our income, and that provide a good guide on how to manage our finances from month to month and in the longer-term. At the end of this, we should have a feasible financial management strategy that, barring totally unexpected events, will have a predictable outcome.

If we have done our calculations correctly and responsibly, the strategy should mean we enjoy a reasonable lifestyle without being sued for bankruptcy. This is little different from the basic task, and overall hope, of the fisheries manager!

COMPONENTS OF FISHERIES MANAGEMENT

Governments spend money on fisheries for purposes other than fisheries management. Grants and subsidies are examples of expenditures which are not due to fisheries management. The same applies to expenditures on navigational aids such as lighthouses, positioning systems, fishing harbours, and search and rescue operations at sea. It is appropriate, therefore, at this point to discuss briefly what kind of activities constitute fisheries management, the cost of which is the focal point of this chapter. Fisheries management essentially comprises the following set of activities:

- Research (biological and economic)
- Formulation, dissemination and implemention of management policy and rules
- Enforcement of management rules.

Any sensible fisheries management must be based on an understanding of the marine environment in which a fish stock is located and an assessment of the size and productivity of the stock at each particular time. Also, since the purpose of fisheries management is to increase the economic return from the fishery, it must be based on a thorough understanding of the economics of the fishing activity and measurement of the relevant economic parameters. To achieve the double aim of preservation of the fish stocks and maximization of the economic return from the fishery, total catch and fishing effort must be constrained either directly or indirectly. Fisheries management techniques are, among other things, methods for constraining either total catch, fishing effort, or both. Enforcement of these constraints is, for obvious reasons, an integral part of the fisheries management activity.

FISHERIES MANAGEMENT AS A PUBLIC GOOD

Fisheries management services exhibit many of the characteristics of public goods. Public goods have two characteristics, non-rivalry in consumption and non-excludability. Non-rivalry in consumption means that two or more persons can use the good simultaneously without interfering with one another's use of the good, while non-excludability means that no one can be excluded from using the good.

As already stated, fisheries management services consist primarily of (i) information generation, (ii) the establishment of fisheries management rules and (iii) the enforcement of these rules. Information, once generated, can be used by anyone without reducing it. Thus, the informational or research part of fisheries management is close to being a pure public good. Very much the same applies to the design of a management system and its enforcement. No-one participating in the industry can be excluded from enjoying the benefits. Also, the benefits are little if at all diminished by the number of participants. Given the public goods characteristics of fisheries management, the market system by itself is ill suited to provide these services. Non-excludability essentially means that it is difficult to charge for the services. Therefore, private industry can hardly survive in this area. Moreover, non-rivalry in use means that it is doubtful that it would be economically appropriate to charge for these services.

APPROPRIATE LEVEL OF INFORMATION COLLECTION AND FISHERY REGULATIONS

It does not necessarily follow from these arguments, however, that the government should actually provide fisheries management services. As pointed out above, it is normally the fishing industry that primarily benefits from fisheries research and management. Therefore, it is in its own interest to ensure that these services are provided. Indeed, there is a presumption, often referred to as the Coase Theorem, that under the appropriate conditions agents will overcome externality problems by negotiating the appropriate collective action, provided this maximizes the overall net benefits. In our view, this argument does not really apply in fisheries. There are two reasons why the industry cannot be expected to undertake the appropriate level of information collection and fishery regulations on its own initiative, even if such activities are beneficial to the industry. First, the industry usually lacks the legal authority to deal with the problems that must be solved in order to make fishery regulations successful. This is most obvious for fish stocks that migrate between the jurisdictions of different states or into the high seas; such problems must be dealt with at the inter-governmental level. Also, the industry typically is not authorized to bar new entrants or to impose restrictions on its members. Such attempts would probably run afoul of competition or anti-trust regulations, and there are indeed examples of such cases from the United States. Second, and more fundamentally, information provision and regulation, exhibit strong public goods characteristics that invite free riding behaviour on behalf of industry members. For these two reasons, imposition of fisheries management and information generation from above may be necessary to overcome the problems of under-provision of these services if their provision were to be left to the industry itself.

Government services are not costless, however. A bureaucracy is needed to deal with regulations, gathering and processing of information, etc. Professionals must be hired to do stock assessment and process information gathered about fish stocks, and the necessary equipment, such as research vessels, must be in place. The cost of these activities must be defrayed somehow.

The sums involved are not trivial. In the Commonwealth fisheries of Australia the expenditure on fisheries management has been estimated at 28 million Australian dollars in 1991/92 while the total value of landings was about ten times that (273 million). Fisheries management costs in the United Kingdom are reported to have been about 45 million pounds in 1996/97, which corresponds to about 7.5 per cent of the value of all landings of fish (Hatcher and Pascoe. Government expenditure in the United States on fisheries management amounts to 15 per cent of the gross revenue. The expenditure on fisheries management in Norway has been about 8 per cent of the value of landings in recent years while in Iceland it has been about 3 per cent, and in Newfoundland 15-25 per cent. Even if the definition of costs, undoubtedly, is not identical in these studies, if only because of different accounting practices across countries, these figures clearly show that (i) fisheries management costs are high and (ii) vary substantially from one country to another.

COST RECOVERY FROM INDUSTRY

We suggest three important reasons why cost recovery from the fishing industry has been moving up on the public agenda in several countries. First, the increasing financial burden of various social obligations undertaken by governments in industrially developed countries, combined with the reluctance among the general public to pay higher taxes, has been putting government finances under an increasing strain. This has given rise to an increasingly critical scrutiny of the public services that are being provided and how they are being financed. Second, related to the first reason, is the public demand for more efficiency in production in general. The incentive effects associated with paying for public services from general government revenue may be undesirable, as there is no link between the value of the services provided and the cost of providing them.

If those who benefit from a particular activity are also the ones who pay for it there would be a link between the value of these activities and their cost. Establishing such a link, therefore, is likely to enhance economic efficiency by providing incentives to determine whether and to what extent a particular service is worthwhile. The third reason for the increasing interest in management cost recovery from the fishing industry is the growing realization that many fisheries, far from being a subsistence industry, can actually yield substantial economic rents, provided they are properly managed. Therefore, there is no need to subsidize them by providing them with management services free of charge. In fact, due to the common property problem, such supports may even be counterproductive. From this perspective, it is no surprise that cost recovery is furthest along in fisheries where an efficient fisheries management system has already been adopted.

As far as efficiency is concerned, the question of whether management costs should be recovered from the industry may be said to have two major aspects:

Benefits of Fisheries Management Materialize

The benefits of fisheries management materialize in the form of a greater net production of goods and services. This need not mean a greater production of fish. Economic overexploitation means that too much manpower, equipment and other means of production have found their way to the fishing industry. Sometimes, perhaps often, this means that the long run productivity of fish stocks has been harmed to the extent that more fish could be caught in the long run by using less of these factors of production. But economic overfishing always means that in the long run more of goods and services other than fish can be produced by reducing the input of manpower, capital, etc., in the fishery and directing it elsewhere. However, in the short run there may not be much economic gain, because of immobility of labour and capital.

The general level of taxes, and how they are levied, has an impact on the real side of the economy. Both direct taxes on incomes, and indirect taxes such as the value added tax, affect work effort and whether it is directed to production of marketed goods and services or to activities out of reach of recorded market transactions. Taxes on the products of specific industries, or income generated in such industries, affect the allocation of resources by making such industries less profitable than otherwise, thus reducing the level of production in such industries. Would it be better, from the point of view of optimal taxation, to have the fishing industry pay its management costs? Defraying these costs from general government revenue adds to the general burden of taxation and the deadweight loss associated with that taxation. Having the industry pay these costs through a tax on incomes in the industry or a levy on its production would make it less profitable and discourage the use of factors of production in the industry.

In fisheries which are imperfectly managed and where there is still a tendency to use too much manpower and capital this would be a desirable effect, as it would mitigate against excessive use of factors of production. Consider, for example, the case where a purely "biological" management regime is introduced, for the purpose of achieving maximum sustainable yield. One way of keeping total landings within the maximum sustainable yield limit, used, for example, in the United States and Canada, is to stop the fishery when that limit has been reached. Rebuilding fish stocks through such management would eventually make the industry more profitable, but without any entry regulations the industry would expand, until it would once more break even. This expansion would amount to an increase in the real costs of the industry in the form of excessive fishing gear, investment in unnecessary boats, etc., and a waste of resources for the economy as a whole. If the industry were to pay for the cost of management it would incur financial costs that would replace some of the real costs it would otherwise incur before it reaches open access equilibrium. Having the industry pay the costs of management would thus save some real resources for society.

Curiously, perhaps, this argument for having the industry pay the management costs becomes weaker or evaporates altogether when the industry is regulated by individual transferable quotas. In this case the limit on the total catch would ensure resource conservation while the transferability of quotas would provide incentives for maximizing the value of the catch and minimizing the cost of taking it. Cost recovery from the industry would have no bearing on the size of the industry; entry into the industry would be effectively closed by the quota system itself.

Yet there is another argument favoring cost recovery in ITQ fisheries; in an economically efficient industry rent would be generated, the amount of which would depend on the productivity of the stock, efficiency of the fleet, and the market price of fish, besides other factors. If designed appropriately, cost recovery would only reduce the rent captured by the firm, without having any effect on the economic efficiency in the industry. From the point of view of optimal taxation this would constitute a gain; the tax burden on the general public could be reduced with no detrimental real effects on the fishing industry.

Incentive Effects of Cost Recovery

It has already been mentioned that paying for government services out of general government revenue provides a weak link or no link at all between what these services are worth to the user and what they cost. This leads to well known problems of the appropriate supply of public services for which the user does not have to pay. Excess demand for fisheries management services is perhaps not particularly likely to come from the industry even if they are free of charge; in fact some industry participants tend to see at least some of these functions as an unwelcome intrusion into their business practices. The drive for increased fisheries management may be more likely to come from government ministries and agencies responsible for management, with the restraint being provided by ministries of finance concerned with the overall government budget.

Having the industry pay for fisheries management would establish a link between what fisheries management is worth to the industry and what it costs. Clearly, the industry would be interested in maximizing the value of the services that it gets for its money and in minimizing what it pays for any given amount of services. It is likely that the industry would demand, and in all probability get, influence over management decisions if it would have to pay the management costs.

This raises the question of compatibility between the interests of the industry and the interests of society at large. It has already been noted that fisheries management is a public good, benefiting the industry as a whole and the society of which the industry is a part. But is there a perfect harmony between the interests of the industry and those of society at large? Does the public derive benefits from fisheries management which are distinct from what the industry derives? It goes without saying that the industry will only be interested in management services that benefit the industry directly. Having the industry pay the full cost of management and granting it a commensurate influence over how the money is spent will therefore lead to negligence of management activities that are of benefit to the public at large but of little or no benefit to the industry.

INDUSTRY COMPATIBLE WITH THE PUBLIC INTEREST IN MARINE RESEARCH

Marine research is to fisheries what geological research is to petroleum and other mineral industries. It can be argued that marine research has utility far beyond the fishing industry. Nonetheless, the marine research typically funded by ministries responsible for fisheries and their subordinate institutions is typically targeted for the benefit of the fishing industry. There seems little doubt that the industry has an incentive to contribute to that kind of research, but even marine research that is designed to meet the needs of the fishing industry may have benefits for a larger audience as well. It can be argued, therefore, that this kind of research would be underprovided if the industry alone were to be responsible for it.

Stock assessment is a kind of marine research that is specifically intended to benefit the fishing industry by providing estimates of current stock abundance and making it easier to evaluate the consequences of today's catches on the future availability of fish. This type of research, however, is of limited direct benefit to the public at large. Provided the industry takes a long term interest in its resource base it would have a strong interest in accurate and timely stock assessments.

Therefore, having the industry pay for stock assessment is likely to provide incentives for providing such assessments at the lowest possible cost and to focus research priorities on whatever is most useful for the industry. The industry itself might even take an active part in the stock assessment process by letting its own vessels take samples of fish, fish at random for assessment purposes, etc. In fact such activities have become routine in Iceland, where parts of the fishing fleet conduct random sampling under the direction of fisheries biologists for short periods every year.

In order to ensure efficient utilization of the resources various regulations are necessary; setting of total allowable catches when appropriate, regulations of gear design (*e.g.*, mesh size), protection of juveniles and spawning stock, etc. As for stock assessment, this is potentially an area where "user says, user pays" would have a beneficial effect. Assuming that firms have a sufficiently long-term perspective, or equivalently a rate of time discount that is not excessive, it is in the interest of the industry to put in place regulations necessary for ensuring that the growth potential of the fish is utilized optimally and that the present catch of fish does not diminish future productivity of the stock unduly. A plethora of fisheries regulations now in place in various countries are put in place for a quite different purpose, however. There are regulations that stipulate the size and design of boats, which gear may be used where, what type of boats may fish where, etc. Such regulations more often than not serve no purpose of efficiency; it would seem immaterial whether a fish is pulled up on a hook or caught in a net, except that some types of gear are more selective than others and hence use the growth potential of the fish more efficiently.

Neither would it seem important, as far as efficient use of resources is concerned, whether a fish is taken by a small or a big boat. Such regulations may be of use in limiting controversies among different parts of the industry, but often they are just a way of favoring one industry group against another. While it would seem to be in the overall interest of the industry to eliminate such regulations, their very existence is often a sign of an industry fragmentation and controversy that would prohibit it from acting in concert, which adds an additional reason why governments need to be involved in fisheries management and detracts from the relevance of the Coase Theorem.

Surveillance and enforcement are obviously needed to make any regulation effective. It has been argued that the industry should not pay the cost of such undertakings, any more than other industries pay for law enforcement on land; law enforcement being a collective good to be provided by the state. In our view this argument misses the point. The purpose of commercial fishing is to maximize the net economic benefits from the fishery. It follows that the overriding (and perhaps the only) criterion for fisheries management payment arrangements is that it be as economically non-distorting as possible.

Recording landings is a part of surveillance whenever there are limits to how much can be caught, whether it be at the aggregate or vessel level. Monitoring landings serves a wider purpose, however. Catch data, both aggregate landings and samples derived from these for age structure analysis, are an important input for stock assessment and usually play a major role therein, except for direct acoustic methods. Accurate landings data may therefore be crucial for an accurate stock assessment. Indeed, one of the criticisms levied at management by catch quotas is that it distorts landings data, due to the incentives to highgrade and to underreport. It is, therefore, strongly in the collective interest of the industry to collect accurate and timely landings statistics. In some fisheries in Canada fishermen have been willing to pay for supervision of landings, recognizing the collective good character of this activity and the incentive of each individual in a quota managed fishery to cheat.

Some fisheries management costs may not be due to the activities of the industry but to competing claims on fish resources. Interference of marine mammals with the catches of fish, and a desire on behalf of some groups and governments to preserve these animals, have in recent times given rise to a number of regulations. It is conceivable that preservation of marine mammals and other environmental demands will impose further constraints on the activities of fishing fleets and how much fish they will be allowed to catch. The benefits of these constraints are external to the industry while such preservation would impose costs on the industry in the form of catchable fish being eaten by seals or sea birds and in the form of costly fishing gear that avoids the incidental catching of these animals.

To the extent such claims are deemed legitimate, the role of the industry as a rule maker must be more proscribed than otherwise, with fisheries management becoming a public concern for a much broader range of players than those who are active in the industry. To the extent the industry would be obliged to pay for this kind of management such payment would amount to a special tax on the industry earmarked for environmental benefits for the public at large, rather than a payment for privileges the industry enjoys.

Similarly, other non-use benefits of marine resources, and the management necessary to safeguard the resources for these purposes, are of no concern, and are possibly detrimental to, the industry. An example of such a non-use value is the viewing of fish as wildlife, for which purpose areas have been set aside as marine parks. Having the fishing industry pay for such activities would again amount to a special tax on the industry for the benefit of other user groups. In fact, it appears that management of marine parks could be taken care of by government bodies distinct from those dealing with commercial fisheries, as is done, for example, in Australia.

EVOLUTION OF VIEWS OF FISHERY MANAGEMENT

Fishing is an old activity, and concern about its effects also is old. As early as the fourteenth century, people in England were worried about fishing with a *wondyrchaum* (a fine-meshed trawl), which they feared was killing enormous numbers of small fish. By 1716, minimum mesh sizes and minimum size limits for various fish species were in effect. Many forms of fishery management have evolved based on traditional knowledge gained by fishing peoples. They included cultural practices, community agreements, and government controls. But in general, and despite early regulations, management of marine fisheries was minimal before the middle of the twentieth century, with a few notable exceptions such as the International Pacific Halibut Commission.

Although there has long been concern over the effects of fishing, that concern—like fishing itself—was largely confined to coastal waters until recently. The famous scientist Thomas Huxley expressed his opinion in 1883 that probably "all the great sea-fisheries are inexhaustible", although some great sea-fisheries were already depleted by then.

The Atlantic halibut fishery had collapsed by the early 1880s and has not yet recovered. And the U.S. Fish Commission had been established a dozen years earlier (1871) to find the causes of declining New England fisheries. By 1919, W.F. Thompson (1919) recognized that sea fisheries were not inexhaustible, but he identified a difficulty that remains critical to this day: "Proof that seeks to modify the ways of commerce or of sport must be overwhelming."

He developed research to concentrate on what "is necessary to the perpetuation and prosperity of the fishery." To protect the Pacific halibut (*Hippoglossus stenolepis*) from the fate of the Atlantic halibut, the International (United States and Canada) Pacific Halibut Commission was established in 1924 with "a competent man as director of investigations": W.F. Thompson. While and before Thompson was investigating the North Pacific halibut fishery, Heincke was developing catch-curve analyses as a basis for proposed minimum size limits for North Sea plaice (*Pleuronectes platessa*) and demonstrating racial differences in North Sea herring. The International Council for the Exploration of the Seas was formed in 1902, motivated by the desire to understand and predict fluctuations in fish stocks.

North Sea fish increased in size and numbers during both World Wars I and II, because naval action forced a reduction in fishing. That phenomenon led to wider recognition that fishing did affect fish stocks and that the effects were at least partly reversible. Those observations led Graham to explain how increases in fishing power allowed catches to be maintained or increased even when fish stocks declined, leading to a waste of time and money by comparison with fishing at a maximum sustained yield. Gordon (1954) provided the economic theory to support Graham's observations.

In the 1950s and 1960s, scientific methods of stock assessment and estimation of fishery yields were developed based on growth rates and age composition of the catch. At the same time, fishing vessels based many hundreds or thousands of miles from fishing grounds ("distant-water fleets") changed the face of fishery management, leading to exclusive economic zones, international treaties, international fishery-management bodies, and continued international disputes about fishing. One of those international bodies, the U.S.-Canadian International Commission for Northwest Atlantic Fisheries, implemented an early attempt at incorporating ecosystem management into fisheries. When it set catch limits for a particular species, it took into account the expected bycatch of that species in fisheries for other species. It also concluded that the total sustainable catch rate for all species was less than the sum of sustainable catch rates for the individual species.

Many factors have reduced the effectiveness of management and prevented the adoption of scientific advice. In a few cases the advice itself was not correct. Progress in developing better scientific perspectives for management and more equitable and effective ways of implementing management goals continues, but it has been slow. In some cases the obstacles seem to be overwhelming. Significant societal, commercial, and governmental economic incentives and pressures often lead to unsustainable fishing practices. It is difficult, for example, for a poor, hungry family to stop or even reduce fishing if its basic food needs depend on fishing and it has no resources with which to supplement reduced food from fishing. In addition, most such families live in countries whose governments have limited resources to apply to the problem. In wealthier societies, economic pressures can be as great as the basic need for food. But it is clear that the long-term costs of overexploitation of fishery resources are even greater than the short-term costs of reducing catches. Options for achieving sustainable fishing are the topic of this report.

VIEWS OF FISHERY MANAGEMENT EVOLVE

As views of fishery management evolve, the importance of terms becomes apparent. For example, it is common to refer to fishing as "harvesting the resource." But the term *harvest* usually includes the idea that an investment has been made in a crop, which is not true of most marine-capture fisheries. Another term often used is *underutilized*, which implies that greater exploitation of a particular fish stock is desirable. This term probably reflects one purpose of many fishery-management institutions—that is to promote fisheries. In this report those terms and some others are avoided; in general, the committee has sought to use the terms *catch* or *landings* instead of *harvest* and *lightly fished* or *subject to low fishing mortality* rather than *underutilized*. The term *stock* might be linked by analogy to stock of a commodity. It also is difficult to define biologically, and so the committee has used the term *population* where possible. It is extremely difficult to find terms that do not express any societal values. Thoughtful evaluation of the values implicit in many terms and replacement of some of them could be an important part of achieving sustainable fishery management.

Report Organization

The stage by reviewing information on the state of marine fisheries. The committee attempted to be comprehensive although not exhaustive. Thus, enough information is presented for a general overview, but not all relevant. The effects on ecosystems of fishing and of environmental changes. What is known about the factors that contribute to the conditions.

These include incomplete scientific information, scientific errors, failure to heed scientific advice, conflicting and unresolved goals and values, failure to consider all ecosystem components, institutional failures, social and economic incentives that do not favour sustainable resource use, changes in perspectives, new information, and perhaps a failure to recognize the limits of science.

The potential usefulness and practical application of an ecosystem-based approach, alternative institutional structures, social and economic arrangements that hold promise for improving sustainable fishing, marine protected areas, scientific questions, and research needs. The emerging recognition that marine ecosystems have values in addition to their production of food and the importance of that recognition in developing ecosystem-based approaches to sustainable fishery management.

DEGREE OF FISH-STOCK UTILIZATION

FAO periodically reports the degree of utilization of global fish stocks, classifying fisheries as underexploited, moderately exploited, heavily to fully exploited, overexploited, depleted, and recovering. The largest number of fisheries (44 per cent) are classified as heavily to fully exploited. Twenty-five per cent of stocks have been fished beyond sustainable limits (overexploited, depleted, and recovering).

For the United States during the period 1992–1994, the picture was similar despite slight differences in terminology: 12 per cent of 275 stock groups were classified as underutilized, 34 per cent as fully utilized, 23 per cent as overutilized, and 31 per cent were of unknown status. Of the 191 stock groups whose status was known, 82 per cent were fully utilized or overutilized. A U.S. example of a formerly overexploited and now recovered stock is striped bass (*Morone saxatilis*): Georges Bank haddock (*Melanogrammus aeglefinus*) represents a depleted stock.

Globally, some increase in exploitation might be possible for 32 per cent of the landed species, but Garcia and Newton (1997) noted that, given past experience, heavily to fully exploited fisheries are likely candidates for future overfishing. This assertion is demonstrated by Alverson *et al.* (1994), who reported (based on FAO data) that from 1980 to 1990 the number of overexploited fisheries increased by 250 per cent, whereas the number of underexploited fisheries decreased by about 75 per cent. Depleted species are being replaced in today's catches by species that were less heavily fished in the past.

For example, the Chilean Inca scad (*Trachurus murphyi*), Japanese pilchard (*Sardinops sagax melanosticus*), South American pilchard or Chilean sardine (*Sardinops sagax sagax*), and skipjack tuna (*Katsuwonus pelamis*) replaced chub mackerel (*Scomber japonicus*), Atlantic mackerel (*Scomber scombrus*), Atlantic cutlassfish (*Trichiurus lepturus*), and saithe (Atlantic pollock, *Pollachius virens*) in the top-10 species list between 1973 and 1993. In the United States, skates and dogfish have replaced more commercially valuable fish on Georges Bank. This process of depletion of one resource and replacement by another is limited by the number of potentially catchable and usable species; depleted species may not return to previous abundance levels. For example, Atlantic halibut (*Hipploglossus hippoglossus*) and spring-spawning Icelandic herring (*Clupea harengus*) have not recovered from overfishing, although they probably would if mortality were reduced. Many shark populations appear to be declining as well. For instance, Van der Elst (1979) described the impact of South African antishark nets on local populations of oceanic sharks, and the resultant and unanticipated decline in nearshore bony fishes. On a global scale, Manire and Gruber (1990) concluded that sharks were overfished by approximately 30 per cent per year in U.S. waters; and cited domestic demand for shark meat; wasteful fisheries practices, especially discarded bycatch of sharks; irrational dread; and an increasing global demand for shark fins as major factors contributing to excess fishing mortality of sharks.

In addition, unexploited fish populations that are long lived and slow growing cannot support high exploitation rates, unlike populations of faster-growing, short-lived species. For example, the five species of the genus *Sebastes*, including the Pacific Ocean perch itself (*S. alutus*) and the northern (*S.*

polyspinus), rougheye (*S. aleutianus*), sharpchin (*S. zacentrus*), and shortraker (*S. borealis*) rockfishes off the northwestern and Alaskan coasts of the United States and the coast of British Columbia, are all slow-growing and were severely overfished, although they have now largely recovered. The marbled rockcod (*Notothenia rossi*) in the Southern Ocean also has been severely overfished.

Catch Per Ton of Fishing Vessel

Approximately 3.5 million vessels are engaged in fisheries worldwide; about two-thirds are small undocked vessels, but the total also includes about 24,000 high-seas fishing vessels of more than 500 gross tons. The gross tonnage of the world's fishing fleets (decked vessels only) increased by an average of 2.9 per cent annually from 1970 to 1992. This rate of increase in gross tonnage exceeded the rate of increase in catch (1.8 per cent annually) during the same period: the ratio of metric tons of fish caught per ton of fishing vessel decreased from 4.3 in 1970 to 3.0 in 1992.

Assuming that additional fishing capacity has at least the same average fishing power per ton as the pre-existing fleet—almost certainly true—and that the new vessels are used at least as much as the older ones—probably the case—the decreased ratio of catch to fishing tonnage provides further evidence that fish populations declined on average during this period. The discrepancy between fishing capacity and catch is even greater when one considers the increase in fishing power of vessels as a result of technological improvements. Based on an analysis conducted by Fitzpatrick (1995), the rate of increase in fishing power resulting from technological improvements has averaged 4.4 per cent annually since 1965. Garcia and Newton (1997) fit a production model to global catch and gross tonnage data, adjusted for fishing power increases. Their analysis indicates that catch is higher than the maximum sustainable yield of world fisheries and that fishing capacity is too large to be economically efficient.

The decline in the per-ton catch rate of fishing vessels also indicates an economic problem, although lower catches have probably been partially offset by price increases. The economic problems also include the substantial debt service or depreciation of fishing vessels. Government subsidies have been used worldwide to increase employment and food supply. Subsidies have probably stimulated excess growth in the world's fishing fleet and must be a major factor in poor economic performance. They may amount to as much as $27 billion per year, although information about subsidies and how people and organizations react to them is not readily available.

UNITED STATES OVERVIEW

Fishing Sectors

Marine fishing activities in the United States are divided among commercial, recreational, subsistence, and indigenous sectors. The balance of activity among these sectors depends on the areas and species fished and whether the comparison is made in terms of weight or number of fish landed or dollars injected into the U.S. economy. All sectors are subject to fishery management in the United States through the regional fishery management councils and, in some cases, through state and international agreements as well. Fisheries are important to the culture and social structure of their practitioners and can have a major economic impact, at least regionally.

Commercial Fisheries

The United States has the largest exclusive economic zone (EEZ) of any nation, covering about 11 million km^2.The United States was the fifth-largest fish producer in 1993, following China, Japan, Peru, and Chile. The first-sale value of U.S. commercial landings (4.47 million t^6) in 1997 was estimated at $3.5 billion, with a direct contribution to the gross domestic product (GDP) of $20.2 billion. The United States is also one of the world's largest fish-trading nations, with a deficit of $4.6 billion in 1994 resulting from $12 billion in imports and $7.4 billion in exports.

U.S. commercial landings were relatively stable at about 2 million per year from 1935 until 1977, when the United States extended its jurisdiction over fisheries to 200 miles from the coast and increasingly excluded foreign vessels. At present foreign fishing is not permitted in the U.S. EEZ, although in some cases—for example, menhaden (*Brevoortia tyrannus*) in the Gulf of Maine—foreign processor vessels receive catches from the U.S. EEZ. Since 1977, landings have more than doubled, to 4.47 million t in 1997. The rapid rise in U.S. catch in the late 1980s was due primarily to the walleye pollock fishery that resulted from displacements of foreign vessels during the 1970s and into the 1980s. About half of the U.S. landings are from the fishing grounds off Alaska, primarily walleye pollock (*Theragra chalcogramma*). Pacific cod (*Gadus macrocephalus*), and various salmon (*Oncorhynchus*) species. As is true for most fishing nations, U.S. fishers are dependent on a small number of species, with almost 50 per cent of the catch composed of walleye pollock from the Pacific Ocean and menhaden (*Brevoortia tyrannus* and *B. patronus*) from the Gulf of Mexico and Atlantic Ocean.

Recreational Fisheries

Recreational fishing also is important in the United States. Although the recreational catch is only about 2 per cent as large as commercial landings for all species combined (90,000 t in 1994), there are more than 17 million marine recreational fishers, who in recent years made more than 66 million fishing trips per year, caught about 360 million fish, and spent $25.3 billion per year on fishing-related activities, comparable to the contribution to the GDP of commercial fisheries. For some fisheries in which both commercial and recreational fishers participate (*e.g.*, summer flounder [*Paralichthys dentatus*] and bluefish [*Pomatomus saltatrix*]), the recreational catch is a significant portion or even a majority of the total.

Recreational and commercial fishers often conflict over management goals and methods for various fisheries. In some cases, recreational fishers are effective at influencing policy, as for example recent restrictions they supported on the use of nets in coastal waters of various states (including a legislative ban on gillnets in Texas in 1988: California's Proposition 132, which banned net fishing starting in 1990: a Florida legislative ban on coastal nets that passed in 1993: and a Louisiana legislative restriction on nets passed in 1994). In other cases, they are not successful. The allocation of available marine fisheries resources between commercial and recreational sectors is a major issue for regional fishery management councils and in the political arena. Some of the disputes and the differences—and occasional agreements—between commercial and recreational fishers are described in almost every issue of *National Fisherman* and *Saltwater Sportsman*. The resolution of such disputes and allocation controversies is made more difficult because recreational landings often are underreported or not surveyed. Serious allocation disputes have been limited thus far primarily

INDIGENOUS PEOPLE'S FISHERIES

Indigenous people's fisheries are a minor part of total catches but are particularly important in cultural and social terms. Indigenous marine fisheries in the United States—primarily in Washington, Oregon, California, and Alaska—are subject to treaties between the United States and tribal groups. Tribal fisheries for salmon include commercial, ceremonial, and subsistence uses. The Northwest Indian Fisheries Commission handles treaty rights related to salmon in the Puget Sound area.

The Columbia River Inter-Tribal Fisheries Commission represents four tribes in the Columbia River basin of Oregon. Fishing by three tribes in the Klamath River Basin in California is not protected by treaty, but 50 per cent of Klamath River chinook salmon are allocated to these tribes by government regulation. The Pacific Fisheries Management Council, as well as its Scientific and Technical Committee and its Salmon Technical Team, have Native American tribal representatives. There is also a Native American allocation for sablefish (*Anoplopoma fimbria*) off the coast of Washington. For communities in western Alaska, which are largely populated by Alaska natives, Section 305 of the Magnuson Fishery and Conservation Act and Section 111 of the 1996 amendments provide for a Community Development Quota Programme. The programme allots varying percentages of the total allowable catch (TAC) of several fisheries to these communities; by 1999 those communities will be allotted 7.5 per cent of the TAC of Bering Sea groundfish and crabs. The amendments also allow the establishment of a similar programme in the Western Pacific Regional Fishery Management Area (Hawaii and other U.S. Pacific islands).

Subsistence Fisheries

Subsistence fisheries—fisheries conducted for food, material, and fuel but not primarily for commerce or recreation—occur in many parts of the world, most commonly in non-industrialized and tribal societies. Although the sale of fish for cash is not included in subsistence fishing, trading fish for other food or services is an important part of subsistence economies in many places, and cash from activities in market economies is used to finance subsistence fishing. Subsistence fishers, like others but perhaps to a greater degree, develop a large store of traditional knowledge. Subsistence fishing is recognized in many laws and regulations. It does not usually constitute a major portion of the landings except locally.

Status of U.S. Fisheries

In its most recent assessment of the condition of U.S. fisheries, NMFS (1996a) evaluated 275 stocks caught by fishers in nearshore coastal waters, the EEZ, and the high seas beyond the EEZ for the period 1992–1994. Of the 191 stocks for which information was available, 33 per cent were overutilized and 49 per cent were fully utilized, leaving only 18 per cent underutilized. Forty-six per cent were below the level of abundance required to produce the greatest long-term potential yield. The long-term potential yield of the U.S. fisheries within the U.S. EEZ is estimated on a single-species basis to be 8.1 million t per year, which is much greater than the recent yields. Based on the calculations in that estimate, for the United States to achieve its potential increase in long-term potential yield, some ("underutilized") fisheries would need to be fished more heavily, but, more importantly, fishing on overutilized stocks, bycatch, and unaccounted mortality will need to be reduced so that stocks can rebuild. The estimated long-term potential yield and maximum sustainable-yield levels can be used as reference points to help guide the sustainable development and prosecution of fisheries or the rebuilding of marine fish stocks that have been overfished.

Although there is limited information available regarding the overall economic performance of U.S. fisheries, they are undoubtedly suffering from over-capitalization at a national level (with some regional exceptions), as has been reported by FAO for fisheries worldwide. NMFS (1996b) reported that there were about 23,000 commercial fishing vessels in the United States in 1987 (the latest year for which there is good information), which is more capacity than is required to achieve the long-term potential yield from U.S. fisheries. For example, the capacity off Alaska has been estimated to be two and a half times that necessary to catch the available resources (North Pacific Fishery Management Council [NPFMC] 1992). Major losses in revenue from New England fisheries have resulted from overfishing, driven in large part by excess capacity.

A Canadian Example: Northern Cod

The case of the northern cod is an example of the effects of overfishing as well as institutional difficulties in applying scientific findings to management. This overfishing occurred despite reasonably conservative target fishing mortalities; the problem was largely due to systematic errors in stock assessments exacerbated by unreported (illegal) discarding of small fish and perhaps unreported catches, and later to a failure of management to respond quickly to corrected assessments.

The fisheries of the Atlantic Canada region have been dominated by groundfish; the cod fishery was unquestionably of greatest importance. Cod (*Gadus morhua*) served as the base of the fishing industry in Newfoundland, Nova Scotia, and other provinces in the region (*e.g.*, New Brunswick). The northern cod fishery is an instructive example of overexploitation of a fishery. It has been much discussed, recently by Walters and Maguire (1996) and Hutchings and Myers (1994), who focused on fishery biology, and by Neis (1992), Steele *et al.* (1992), and Finlayson (1994), who focused on the sociology of science. A combination of lack of data, improper handling of available data, and overconfidence in methods led to overfishing and the collapse of the fishery. The offshore catch of northern cod expanded from approximately 240,000 t annually in the mid-1950s to a peak of 700,000 t in 1968. Total catches of northern cod declined steadily thereafter. By the early 1970s, distress in the northern cod inshore fishery also was evident.

The Canadian government planned to rebuild the resource through reduced fishing by the distant water fleets. As the resource rebuilt, the total allowable catch (TAC) for northern cod would gradually be increased. The management strategy adopted was expected to result in sustainable catches averaging 20 per cent of the exploitable biomass by the late 1980s—roughly 400,000 t annually. In the years immediately following implementation of this plan, the northern cod resource appeared to be rebuilding as planned, but the actual landings never achieved even 260,000 t per year. The offshore sector of the cod fishery always succeeded in taking its allocation, but the inshore sector landings declined by 35 per cent from 1982 to 1986.

In response to these trends, the Canadian Atlantic Fisheries Scientific Advisory ommittee (CAFSAC) undertook in 1988 a review of its stock-assessment methods. CAFSAC concluded that the northern cod stock was in fact substantially smaller than previously believed, a view confirmed by an independent Northern Cod Review Panel, which estimated that the actual fishing mortality rates had been at least double those projected in the Canadian management strategy. CAFSAC concluded that the northern cod TAC for 1989 should be reduced from the continuing 266,000 t to 125,000 t. The Canadian government, fearful of the economic disruption and dislocation that such a draconian reduction in the TAC would entail, reduced it only to 235,000 t. In mid-1992, after poor catches, a moratorium was established on all directed commercial fishing for northern cod for a period of two years, during which deterioration of the stock continued. The moratorium was extended, remains in place, and is expected to remain in effect for the indefinite future. Recent reports indicate that the northern cod stock is at a historically low level and that there are, as yet, no significant signs of recovery of the stock.

The causes of the resource management catastrophe are the focus of intense debate in Canada, although the proximate cause is clearly overfishing supported by erroneous assessments of stock size and fishing mortality. The reasons for the errors in stock assessments are complex. Overcapitalization in the fishery may have exerted pressure to interpret stock-assessment data in an excessively optimistic manner, as did an overreliance on the science and culture of quantitative stock assessment.

The northern cod stock-assessment procedures appear to have been flawed from 1977 until at least 1985, owing to statistical inadequacies of the biomass model used, overreliance on catch-per-unit-effort data, variability of the data set, and relatively short and unreliable data series. Ironically, the northern cod is one of the few examples that seems to show a clear and positive relationship between parent stock and recruitment. Although the intuitive expectation is that the more spawning adults there are in the population, the more recruits there will be, most fish populations do not show such a relationship. For the northern cod, estimates of recruitment were not corrected for changes in spawner biomass, which themselves were overestimated. Fishing mortalities were underestimated, probably because of unreported discards of young fish—a significant source of mortality—and perhaps unreported or underreported catches of adult fish.

Recent analyses indicate that the cod populations in the western Atlantic are not the only ones in danger of being overfished. For example, Cook *et al.* (1997) concluded that there is an urgent need to reduce the exploitation rate on North Sea cod to avoid risk of collapse (they also found a significant relationship between parent stock and recruitment). Several NMFS assessments of cod in the Gulf of Maine also reached this conclusion, recently confirmed by the National Research Council.

TRENDS IN FISHERIES MANAGEMENT

Poorly defined property rights, increasing effort and decreasing catches are resulting in conflict in fisheries throughout the world. While these effects are common and well documented in highly regulated fisheries in the European Union (conflicts over allocation of quota for example) and North America (for example conflicts over access to salmon stocks between the USA and Canada); the effects upon fisheries in developing countries, while under-researched are invariably more dramatic and can have long-term implications for the promotion of sustainable livelihoods. While there is as yet little evidence of a sharp increase in natural resource conflicts it is apparent, however, that the consequences of conflict in the management of natural resources are becoming more detrimental to long-term sustainable exploitation and are effecting an increasingly large number of people.

Although there is an extensive literature on natural resource conflicts, little work has been done on a) analysing the causes of conflict beyond the case-study arena or b) applying economic tools to the study of conflict. In an attempt to redress the balance, this chapter sets out some initial findings from an on-going study into the management of conflict in tropical fisheries. Drawing on New Institutional Economics and common property resource management theory, it analyses how and why fisheries institutions adapt to changing circumstances and the role of conflict in the process. Using evidence from Ghana it examines the emergence of fisheries management institutions under differing access regimes and analyses the factors which appear to have influenced institutional change.

The paper argues that conflict can be the result of rising transaction costs and the inability of natural resource institutions to manage these changes. It also argues that conflict can be both a positive and negative force and should not necessarily be eliminated altogether. The chapter is divided into four sections.

First, it locates fishing as an economic activity within developing countries and looks at the interactions between local, national and international policy objectives and how these can impact upon the development and exploitation of the resource. In the second section, conflicts are described and why they emerge in natural resources management is discussed. Focusing on the role of institutions it looks at how changes in institutions might affect the management of common property resources.

Section three draws on initial findings from field research conducted in Ghana between March and May 2000. In the final section, the paper presents a number of conclusions as to the process of conflict formation and the possibilities for managing it. The role of fisheries in developing economies Fisheries play an important role in the local and national economies of many developing countries where they provide food, income and employment for large sections of the population. However, the framework within which they operate is often complex and vulnerable to failure. The absence of clearly defined property rights, the trade-offs between national and local needs and the impact of rising pressures on the economy can, however, all contribute to the failure of fisheries to positively contribute to the development process. The national perspective: Following the second world war the rapid expansion of fishing capacity in many developing countries was promoted through the so-called Blue Revolution.

New fishing technologies and fishing methods coupled with a desire to improve nutrition and generate foreign exchange led many countries to build on fishing activity already present. Some of this development was as part of national development programmes, in other cases, international loans and funding programmes helped promote the development of national fishing industries. As a direct result of this drive to build up national fishing capacity, the participation of developing countries in global fishing activity rose dramatically between 1955 and the 1980s. The rewards to be gained from participating in the global fishing market were potentially large, and many benefited, albeit in the short-term, from their forays into fishing. Increased export-revenues and the multiplier effects associated with expanding industries all helped boost the development of many developing coastal states.

The long-term consequences of such developments, have, however, had a dramatic effect. Rising incomes and increased borrowing of cheap petro-dollars contributed to chronic debt problems as the global economy staggered to a halt in the early 1980s; an attitude of positive speculation on fish-stocks, lack of proper assessment and regulation led to the over-exploitation of a number of important stocks (Schurman, 1996; Thorpe et al 2000).

As a consequence of the international debt burden and the strictures placed on the economy by the Structural Adjustment Programmes, trade-offs between local and national needs put further pressure on resources.

The downsizing of the state and the increased role of the market as the resource allocation mechanism saw many developing countries expand their export fisheries to generate revenue whilst at the same time the power to regulate such activity was hampered bythe limited capacity of Fisheries Departments to undertake interventions. The push for export revenue and the renting of grounds to Distant Water Fleets has resulted in spatial conflicts between industrial and artisanal fleets, state support for local fisheries has been constricted and external economic pressures now drive national policy formation.

From a local perspective: In agricultural-based economies, natural resources form the social safety-net that supports large parts of the population. Access to common property natural resources such as timber, fuelwood, grazing land, fisheries, forest products and irrigation water are fundamental to the livelihoods of many of the world's poor. These natural resources are often held as common property and the survival of rural populations has traditionally relied upon a complex set of institutional arrangements that govern use of and access to these resources.

The formation of institutional access and use arrangements is well-documented. Whilst many communities have succeeded in maintaining their common property resources through collective action, in some cases pressures internal and external to the community have caused arrangements to collapse. The outcome of a resource becoming open-access is likely to be over-exploitation and the dissipation of resource rents. In terms of development, then, as the ability of a community of users to exercise rights over their resource is eroded so the role of the resource as a safety-net is undermined.

The pressures that lead to the erosion and collapse of common property resources are diverse. They can result from a rapid rise in population (either through natural growth or an influx of migrants); changes to the ecological status quo (through droughts or flooding, for example); national policy changes that affect rural employment rates; technological change that results in higher harvesting potential and social change that can alter income differentials, systems of property rights and market systems.

An important outcome of the breakdown of common property arrangements is the emergence of conflict between stakeholders. However the precise mechanism is not fully understood. A review of the conflict literature reveals numerous theoretical approaches to describing and explaining conflict.

Conflict is a function of social structure (sociology), of power or class relations (political science) and of individual utility maximisation (economics). It is both positive and negative, constructive and destructive, violent, coercive and non-violent. Thisdiversity of approaches reflects the wide range of disciplines that have addressed the subject. There are many possible definitions of conflict, the following synthesis of definitions is used in this chapter: Conflict is a situation of non-cooperation that involves groups of people with differing goals and objectives and yet Conflict is dynamic and can be a positive catalyst for change Although the term 'conflict' often has negative values attached to it, this is not necessarily the case.

Conflict can be described as negative when the outcomes are a zero-sum game (no-one benefits) rather than a positive sum-game (everyone benefits) or where there is a deadweight loss of social resources as a result of the 'guns vs butter' argument. However, it is also important to bear in mind that conflict can be positive and attempts should never be made to eradicate it completely or to prevent it emerging. Conflict encourages goods to be produced more cheaply, government to become more efficient, flaws in the set-up of institutions to be ironed out and allows society to function efficiently by resolving small conflicts often.

At this point, the question of the difference between conflict and competition arises. Competition implies the existence of rules, conflict arises when those rules are, for whatever reason, disbanded or ignored. In order to help clarify whether conflict is positive or negative it is often helpful to look at what the conflict is over and how fundamental that disagreement is to the social status quo. Aubert (1963), Boulding (1966) and Powelson (1972) all differentiate between conflicts that are 'within consensus' and those that are 'over consensus'. In the former case the parties to the conflict agree about the value of what they seek but not the means of achieving it. In the latter case the parties to the conflict are unable to agree on the value of what they seek nor on how to achieve it.

The potential impact of conflict is thus dictated by the degree of consensual framework within which they are contested and the degree of conflict over basic consensus.

AN ECONOMIC EXPLANATION FOR CONFLICT

Economic theory provides two reasons why conflicts emerge over natural resources:

- The allocation of scarce resources requires trade-offs which become increasingly difficult as the demand for and supply of resources changes and
- Short-term personal gain wins out over long-term social needs or benefits.

The resource allocation issue: Natural resource conflicts occur when the resource in question has become so scarce or degraded as to raise issues of allocation amongst the community of users. Under perfect environmental conditions, as the ratio of users to resource grows so expansion takes place: extra land is brought into cultivation, new areas of forest are exploited, different species are fished or fishery activity moves along the coast or further out to sea. Powelson argues that so long as the answer to 'who gets how much' is resolved by producing more, conflicts over allocation are positive because they encourage growth. When the ecological boundaries of the resource have been reached, further expansion is no longer possible (without adverse consequences) and the division of the metaphorical cake has to change because the option of increasing the size of the cake is no longer possible. At this point, resource allocation mechanisms come under increased pressure to satisfy all users. The equity of resource allocation tends to diminish as the ability to expand production decreases.

As government departments attempt to placate all the stakeholders, so resource allocation decisions have to change and conflict ensues as certain stakeholders gain precedence over others. This is of course nothing new. The trade-offs between equity and efficiency occur in all markets and are faced by all governments. In the developing world however, where democracy may be weak and where powerful interest groups are able to co-opt and unduly influence the decision making process, managing the trade-offs is difficult. Short-term gain vs social needs: There is a large body of literature that uses game theory and behaviour modelling to explain why common property management institutions (in particular) can fail. The Nash Equilibrium and the Prisoner's Dilemma illustrate this point neatly: in the classic decision making model-the Prisoner's Dilemma-the best possible move for player A given the best possible move for player B is to not co-operate.

Thus the socially optimal outcome of the 'game' is not a Nash Equilibrium. McKean (1992:248) notes, in a similar vein that the short term benefits of cooperating are often outweighed by the costs.

Consequently a community opts for the least-cost option which is to withhold their contribution to the collective goal. A third explanation for the emergence of conflict has also arisen from the more recent New Institutional Economics school, this sees institutions and transaction costs as a key element.

From a new institutional economics perspective, institutions exist to minimise transaction costs and are a means of transcending the social welfare dilemmas that arise out of individual action and help maximise collective welfare. They are also the mechanism for dealing with 'gaps' left by market failure in insurance and risk assessment. Transaction costs are literally the costs involved in negotiating a transaction and have been described as the economic equivalent of friction in the world of physics. Transaction costs are the costs associated with gaining information, making decisions and carrying out decisions.

In terms of fisheries, they can be divided into the ex-ante costs of collecting information and making collective decisions in the fishery and the ex-post costs associated with implementing collective decisions. Kuperan et al (1999) argue that transaction costs in fisheries arise, mainly, from the fact that the fishery involve multiple stakeholders and differing objectives and long-term goals. Institutions are important as resource allocation mechanisms-markets and fisheries departments, for example, both have a role to play in allocating scarce resources. What is more, it is now recognised that institutions matter for economic development and recent development research has recognised the need to 'get the institutions right' if effective economic development is to be achieved.

Katterman (1998:294) points out that the OECD has finally recognised that technical co-operation, the long-time philosophical foundation of development practice, can amount to nothing if the institutions that it relies upon for success are weak. Thus for both successful natural resource management and economic development institutions able to adapt to changing circumstances are needed. There are two explanations about why institutions might fail in the face of change.

The first relates to the supply and demand for institutional change, the second to transaction costs. Thomson, Feeny and Oakerson (1992: 132) and Feeny (1988) discuss the supply and demand model for institutional change: when the demand for institutional change (to capture gains not possible under existing arrangements) outstrips the ability to supply change, failure emerges. They list relative factor or product prices, the size of the market, technological change and fundamental decisions of government as causes of change that lead to what they term 'institutional disequilibrium'. However, they also recognise that change depends on the State's willingness and ability to help new institutions emerge.

This is a particularly pertinent issue in the developing world where, as discussed before, state capacity may be underdeveloped. Because in the NIE perspective institutions evolve to minimise transaction costs (through minimising uncertainty)-when it is no longer able to do this effectively its position is weakened and it is increasingly unable to deliver services effectively or be allocatively efficient. Using the NIE paradigm and the supply and demand argument above it could, therefore, be argued that a rise in transaction costs, over and above the institution's ability to accommodate this change leads to inefficient operation. This rise in transaction costs could be due to development pressures (political and economic), environmental scarcity (perceived or otherwise) and structural problems (political and economic) which increase the costs mentioned earlier: the costs of collective decision making, information gathering and collective operation.

In the case of fisheries management, strong and flexible institutions are required at both the national and local level. Arguably, the more nested the structure, the lower the transaction costs between the top and the bottom of the system and the more efficient the institution in its allocation role.

In many developing countries this may not be the case. Communication between the different layers of fisheries management are frustrated, the legitimacy of the ruling body to assign resources is often missing or contested and political factions and the action of rent-seeking elites influence the vertical relationship. Institutions, conflicts and fishing in Ghana The political and economic context

Ghana, the former British colony of the Gold Coast, became the first African colonial state to gained independence in 1957. It has had a number of military coups, but is set, in December 2000 to become the first African state in which a military ruler has handed power to a civilian government through the ballot box. Unlike many other countries in the region, Ghana has experienced comparatively little unrest with its neighbours or amongst its diverse tribal and linguistic groups, and it is perhaps this aura of stability and peace at a national level which influences conditions at a local level. In the 43 years since independence Ghana has made some remarkable advances in development and has also suffered some serious set-backs.

Despite significant advances made in improving some of the standard development indicators (life expectancy has risen by 4 years since 1995; adult literacy has risen by 44 per cent since 1970), Ghana is currently struggling with the dual effects of economic adjustment measures and external shocks. Economic reforms have seen public sector budgets slashed and the impact in the natural resources sector, to name but one, has been devastating. All departments are under funded, and in the fisheries department, this has lead, in particular to a reduced ability to monitor fisheries effectively and enforce regulations. The need to raise revenue has seen the introduction of VAT at 10 per cent which is set to rise to 12.5 per cent later this year. Despite well intentioned publicity campaigns to educate the public as to the need for VAT, this is unlikely to ease the blow of a significant price rise on all VATable goods-particularly on basic grains that have shown sharp price rises since last year. Coupled to this is the rapidly devaluing national currency which, while making Ghanaian exports cheaper, makes servicing the national debt (currently $580 million) more expensive daily and dramatically increases the costs of vital industrial inputs.

Ghana is an oil importer, and recent OPEC price rises have also hit the economy hard. While exports have grown aided by improving terms of trade and export initiatives put in place as a result of economic reforms, these have tended to concentrate on cash crops such as cocoa.

In the context of fisheries, the prevailing economic conditions are having a number of effects. Firstly the day to day costs of fishing are rising. Fuel prices not only impinge on the length of fishing trips, but exchange-rate depreciation has increased the costs of imported inputs such as nets, ropes and outboard motors.

The knock-on effects of such price rises are widespread. Education and health at the local level both suffer as more and more income is diverted away from the 'add-on luxuries' into the basic needs for survival. Although things have not reached crisis level as yet, there is evidence of malnutrition in fishing villages and primary schools lack basic teaching materials (supplied by the community). Despite this catalogue of economic problems, some very interesting results emerged from a recent study of the causes and management of conflict in coastal Ghanaian fishing villages.

MANAGEMENT OF RESERVOIR FISHERIES

The fish production from the reservoirs is low, emphasizing the need for attention to shape and develop the reservoir fisheries from the survey and planning stage to achieve high rate of production and better returns for the fishermen, who represent the weaker section of the society. Majority of these water bodies are not scientifically managed. Only a handful has so far been harnessed on scientific lines, while the others are either half-heartedly managed or even not managed at all.

There are marked variations in the fishery management practices which are followed in various reservoirs within the country. Even though the reservoirs are owned by the Government or Corporate agencies in most of the states, their fishing rights and exploitation systems vary considerably. The fishing systems can be divided into the following broad categories: a). privately owned and managed reservoirs, b). Public water bodies, c). Community water bodies and d). Water bodies managed by the Government. After a scrutiny of the various management practices followed in the country, it is difficult to miss a common underlying spirit of the common property norm. Majority of Indian reservoirs are public properties where a fixed number of licensed fishermen make their living.

The exceptions are the small reservoirs in some states like Karnataka and Uttar Pradesh, which are auctioned to private individuals on annual basis. The following steps are to be taken for the development of reservoir fisheries. Some of them are

1. Pre-impoundment survey of the reservoir
2. Removal of obstacles like tree stumps.
3. Fish farm construction at the dam site for stocking of fish seed adequately.
4. Organization of fishermen co-operative society for harvesting the fish in the reservoir and to take up fish marketing.
5. Implementation of conservation methods to prevent over exploitation and to prevent catching of small fish.

Pre – impoundment survey: Before formation of reservoir studies and investigations were taken up in the rivers connected to know the possible effects of fisheries. In case of the dams constructed, the dam constructing authorities should facilitate necessary fish ways for fish migration. The authorities also assess the fishermen living in the villages and their craft and tackles that are going under the submergence of the reservoir. The above measures are to be taken up before formation of the reservoir to assess and to plan for future development of fishery in the reservoir.

Removal of obstacles under submergence: After formation of the reservoir, it should facilitate easy fishing with modern fishing methods. To facilitate the above, the trees, buildings, boulders are to be removed either by mechanical means or manually.

Construction of fish seed farms at dam site: Some of the reservoirs have got river connection as such, they will have natural stocking of fish seed (auto stoking). The reservoirs, which will have a deep column of water, will greatly affect the breeding grounds of the fish. Further more the height of the dam is another major obstruction for the movement of the fish from the lower stretch of the river to upper stretches.

All these factors leads to low productivity, which drags the attention of many authorities towards the fish seed stocking every year in the reservoirs. Keeping in view of this, planning has to be done for establishment of a fish seed farm near the dam site. The Department of fisheries, Govt. of Andhra Pradesh made an attempt to promoting the fishery wealth in the reservoirs, by constructing the fish seed farms at Thandva, Wyra, Kinnerasani, Palair, Nagarjuna Sagar, Nizam Sagar, Dindi, Jurala, Kadam, Araniyar, Bahuda, Somasila, Sriram Sagar and Lower Manair dam.

Stocking of the fish seed in reservoirs: An adequate sized fish seed should be stocked in the reservoirs. The tender stages of fish seed like fry (22-25 mm), fingerlings (40-50 mm) are susceptible and are to be eaten by the carnivorous fishes. Hence the reservoirs to be stocked with an advanced fingerlings (100-120 mm). Stocking of reservoirs with fingerlings of economically important fast growing species to colonies all the diverse niches of the biotype is one of the necessary prerequisites in reservoir fishery management. This has proved to be a useful tool for developing fisheries potential of such small aquatic systems.

Agarwal recommended the following stocking rates in reservoirs.

1. Large reservoirs (1000-5000 ha) – 500 fry/ha/yr.
2. Medium reservoirs (100-1000 ha) – 1000 fry/ha/yr.
3. Minor reservoirs (10-100 ha)-2000 fry/ha/yr.
4. Small reservoirs (below 10 ha) – 10000 fry/ha/yr.

Srivastava recommended the following stocking rates of fingerlings for reservoirs in India.

1. Large reservoirs (5000-10000 ha) – 200 fry/ha/yr.
2. Medium reservoirs (1000-5000 ha) – 400 fry/ha/yr.
3. Minor reservoirs (up to 1000 ha)-1000 fry/ha/yr.

IMPLEMENTATION OF CONSERVANCY METHODS TO PREVENT OVER EXPLOITATION

It has been observed that fishes of mature major carp fishes migrate to the rivers for breeding, where the fishermen lead to catch them and destroy their breeding grounds. The breeding grounds of these fishes are to be identified and necessary protective measures are to be taken up. It is also observed that during rainy season, the big fishes migrate towards the dam site along the canals and congregate at the gates, which has been catch by the fishermen folk. This is to be reduced.

The conservancy methods like mesh size regulation, observing the closed season for fishing, strict observation of prohibiting the unethical means of fishing are to be implemented strictly by the Government.

Species enhancement: Decline of indigenous fish stocks due to habitat loss, especially that caused by dam construction, is a universal phenomenon. Planting of economically important, fast growing fish from outside with a view to colonising all the diverse niches of the biotope for harvesting maximum sustainable crop from them is species enhancement. It can be just stocking of a new species of introductions. Here introduction means one time or repeated stocking of a species with the objective of establishing its naturalised populations. This widespread management practice has more relevance to larger water bodies where stocking and recapture on a sustainable basis is not feasible.

Introduction of exotics: In India, the fish transferred on trans-basin basis within the geographic boundaries of the country is not considered as exotic and there are no restrictions on them. Thus, catla is not regarded as exotic to peninsular rivers. This is despite the fact that the peninsular rivers have habitats distinctly different from those of Ganga and Bhrahmaputra. Catla, rohu. and mrigal have been stocked in the peninsular reservoirs for many decades now. with varying results. In some of the south Indian reservoirs, they have established breeding populations. The hallmark of the Indian policy on introductions is the heavy dependence on Indian major carps.

There is evidence that the Gagetic major carps have affected the species diversity of peninsular cyprinids. The Indian policy on stocking reservoirs, though not very explicit, disallows the introduction of exotic species into the reservoirs. However, common carp is very popular in reservoirs of the northeast where it enjoys a favourable microclimate and a good market. Silver carp and grass carp are not normally encouraged to be stocked in Indian reservoirs, though they are stocked regularly in a few small reservoirs. The three exotic species brought in clandestinely by the fish farmers, bighead carp..Aorichthys nobilis, Oreochromis niloticus and African catfish Clarius gariepinus have not gained entry into the reservoir ecosystems. They still remain restricted to the culture systems.

Environment enhancement: The improvement of the nutrient status of water by the selective input of fertilizers is a very common management option adopted in intensive aquaculture. If similar environmental enhancement is adopted in small reservoirs, stocks can be maintained at levels higher than the natural carrying capacity of the environment.

However, scientific knowledge to guide the safe application of this type of enhancement and the methods to reverse the environmental degradation, if any. is still inadequate. On account of all these, this is not a very common management tool. China is known to have used this instrument in a big way to augment production from small reservoirs. Cuba, taking a clue from China has tried manuring of small reservoirs using both organic and inorganic fertilizers. This is also practised selectively in the community water bodies of Thailand.

Fertilisation of reservoirs as a means to increase water productivity through abetting plankton growth has not received much attention in India. Multiple use of the water body and the resultant conflict of interests among the various water users are the main factors that prevent the use of this management option. Surprisingly, fertilisation has not been resorted to even in reservoirs which are not used for drinking water and other purposes. Documentation on fertilisation of reservoirs in India is scarce. Sreenivasan and Pillai attempted to improve the plankton productivity of Vidur reservoir by the application of super phosphate with highly encouraging results. As soon as the canal sluice was closed, 500 kg super phosphate with P,O5 content of 16 to 20 per cent was applied in the reservoir when the water spread was 50 ha with a mean depth of 1.67 m.

As an immediate result of fertilisation, phosphate content of water increased from nil to 1.8 ppm and that of soil from 0.242 to 0.328 per cent. Similar improvements in organic carbon nitrogen have been reported from soil and water phases on account of fertilisation. Experiments were also conducted with urea in the same reservoir. Fertilisation can play a key role in many small reservoirs of India which require correction of oligotrophic tendencies. A number of reservoirs in Madhya Pradesh. the Northeast and the Western Ghats receiving drainage from poor catchments show low productivity, necessitating artificial fertilisation. Chinese experience in fertilising the small reservoirs for increasing productivity has been reassuring.

Fertilizers are less effective in soft water with total alkalinity less than 20 ppm. Soft waters have inadequate carbon usually in the form of carbon dioxide and can often be enhanced by applying lime to low alkalinity impounded waters. The application of lime equivalent to 2,000 to 6.000 kg/ha calcium carbonate is generally sufficient to maintain total alkalinity above 20 ppm.

Suggestive Model for Management of Reservoirs

The concept of fishery-business model was developed by Desai in 1984. The main components of this model are macro-fisheries and micro – fishery systems. Macro-fishery system again composed with fish supply system, fish production system and fish marketing system. The micro –fishery system contains fishery input sub system, fishery credit subsystem, fishery extension subsystem, fishery education sub system, fishery habitat sub system, fishery research subsystem, fishery regulation subsystem, fishery administration subsystem, fishery exploitation subsystem, fishery processing subsystem and fishery distribution sub system.

Natural Fishery in Reservoir

The following fishery wealth is available in the major reservoirs of Andhara Pradesh. The Fishes like Wallago attu; Mystus species; Heteropneustes fossilis; Clarias sp; Major carps like – Catla catla; Labeo rohita; Cirhinus mrigala and other carps like C. reba; Thynicthus sandkhol etc. The fresh water Eel, fresh water prawns, Murrels are also available in the reservoirs.

Planning Criteria

In keeping with the need for rapid assessment of the country's small reservoir resources, the following planning criteria are suggested for the resource assessment at the State level for the preparation of an inventory of such small reservoir ecosystems along with the estimates of their potential yields under the categories-a. Reservoirs which are best developed as capture fisheries, b. Reservoirs mostly of local interest having significant potential for fish culture; and c.Reservoirs intermediate in size and potential yield.

These under-utilised fishery resources offer immense scope and potential for generating additional national income of the order of more than ₹ 100 crores per year and providing additional employment to lakhs of fishermen and others through fishing, handling, transport, marketing and ancillary industries. A systematic and integrated approach towards scientific studies and planning criteria for undertaking fish culture in small reservoirs should be so directed as to have an understanding of the following factors.

- The reservoir morphometry and water residence time.
- The physico-chemical characteristics of water and soil.
- The animal and plant inhabitants.
- Growth rate of commercially important fish species, and
- The relation between the inhabitants and the physico-chemical aspects of the environment in terms of population and community dynamics.

It is felt that under the prevailing socio-economic conditions, such short-range studies undertaken for small reservoirs would provide a rapid assessment of their fisheries potential to take up fish culture in them.

RESERVOIR FISHERIES MANAGEMENT IN INDIA

The present low level of fish production in Indian reservoirs can be attributed to inadequate management inasmuch as many of them have high propensities of production from a limno-chemical point of view. In many of the reservoirs, the high rate of the primary and secondary productivity is not channelled to fish production. Insufficient understanding of the reservoir ecosystem often comes in the way for adopting effective management measures. Since fish production from reservoirs is essentially extractive in nature, the essence of management strategy lies in exploitation of natural stocks. Nevertheless, the ecosystem management provides different degrees of freedom for stock maniputation, depending on the size and class of the water body.

One of the possible criteria that can be used to differentiate between capture and culture fisheries is the extent of human intervention in the ecosystem management. While aquaculture systems provide maximum avenues for the man to monitor and change the habitat variables and the biotic communities at will, this freedom attenuates as we proceed from aquaculture to the culture-based and capture fisheries. In a large water body, managed on capture fishery norms, there is little room for altering the habitat variables and the scope for effecting change in biotic communities is limited to stocking and ranching, which have uncertain chances of success. Relative contribution of culture and capture norms in management vary, depending on the category of the reservoir. Medium and large reservoirs are predominantly capture fisheries systems and the management norms are based on the principle of stock manipulation, adjustment in fishing effort, observance of conservation measures and gear selectivity. Selective stocking is resorted to for correcting imbalances in species spectrum and to fill the vacant ecological niches.

The small reservoirs, on the other hand, are generally managed as culture-based capture fisheries, akin to extensive aquaculture, where the main accent is on stocking, fattening and harvesting. An imaginative stocking and harvesting schedule and right species mix hold the key for effective management of small reservoirs.

STOCKING

Inducting fast-growing extraneous species into the ecosystem to colonise the diverse niches is a necessary prerequisite of reservoir management. Since one of the primary aims of stocking is to ensure utilization of the enhanced food reserves, the ideal time to stock new species is the period of trophic burst. Any lapse in this important management measure causes the proliferation of trash fishes by taking advantage of the increased availability of fish food organisms, which in turn, may provide forage base for catfishes. Nagarjunasagar, Tungabhadra, Hirakud and a number of other large reservoirs in India are examples where the minnows, catfishes, murrels and other uneconomic fishes gained grounds in the early years, leading to establishment of long food chains. These reservoirs harbour good standing crops of plankton and benthos, which are not reflected in the fish output. Even intensive stocking at a later stage has failed to reverse the situation.

Since large and medium reservoirs are to be developed on the principles of capture fisheries, it is desirable to stock the species that may breed and ultimately get naturalised in the system through autostocking. This is imperative to meet the long-term objective of obtaining a sustained yield rate. Management involving persistent stocking in large water bodies not only pushes up the input cost, such systems also create many practical difficulties in raising the stocking material in adequate quantities. However, naturalisation of introduced species is quite often beset with many problems. The alien species should find the habitat conducive to its biological and physiological requirements. It should have an edge in the competition for food and finally, the environmental conditions should favour its requirements of spawning and larval development.

Recruitment failure due to the erratic hydrographic conditions that break the breeding rhythm have been found to be the single major factor responsible for the failure of stocked fishes to hold out in reservoir.

In Konar, strong currents wash down the carp eggs into the deep zones of the reservoir leading to their destruction and resultant recruitment failure. But more often, the fishes do not find suitable spawning grounds as in the case of many south Indian reservoirs. Since the breeding of major carps is governed to a great extent by the magnitude of monsoon floods, annual variations in this parameter affect their breeding and recruitment.

Selection of species for stocking

Jhingran has summarized the principles to be followed in the selection of species for stocking as:

1. The species should find the environment suitable for growth and reproduction.
2. It should be quick growing, ensuring high efficiency in food utilization.
3. A fishery comprising herbivores with a short food chain is preferable, as they have a better conversion of primary production to fish flesh.
4. The stocking density should be such that the food resources of the ecosystem are fully utilized and optimum population maintained, consistent with normal growth.
5. The size of the fingerlings to be stocked should be so chosen to get the desired results.
6. Seed should be readily available with minimal transportation cost.
7. Cost of stocking and managing the species must be such that the operation is economically viable.

One of the important considerations is to know the amount of food available in the new environment. This factor has a considerable bearing in determining stocking rates and hence production.

Fish production from unit area is a product of individual growth rate and population density.

From a study of growth rate of various species in a particular body of water, it is possible to assess their optimal stocking density. Efficient utilization of fish food communities increases the carrying capacity and therefore calls for a higher population density which is achieved by addition of species to the original fish populations. In addition, information on differences, if any, in the growth rates of the endemic and introduced species, and the time taken by the introduced species in attaining harvestable size would provide insight into the production dynamics of the system.

Competition

Both intra-and inter-specific competitions are to be considered in the stocking programme. Situations where two or more species use a similar resource, such as food or space, lead to overcrowding and poor growth rate. Under a higher than optimum stocking rate, though production may be high, the individual growth rate will be so small that to attain a marketable size a long growing period will be needed.

On the other hand, if bigger fishes are needed, the rate of stocking should be lowered and a low production will have to be accepted. Similarly, when a marketable size is to be attained in a shorter period, stocking rate will have to be lowered to allow faster growth. Thus, a desired balance among stocking rate, population density and growth is to be maintained with enough flexibility so as to swing it to suit the changes in environmental factors. Such a plan must determine tentative stocking rates and population thinning, depending on the need.

The sizes of fingerlings that are stocked in Indian reservoirs come under the pre-recurit phase and so, up to the size of entering the exploited phase they are prone only to natural mortality. Therefore, a knowledge of the natural mortality rate is essential as due compensation can be provided for it while computing optimum stocking rates. Jhingran and Natarajan, while evaluating the stocking rates of Damodar Valley Corporation (DVC) reservoirs, arbitrarily recommended a compensation factor for natural mortality @ 25 per cent for reservoirs with no large predaceous fishes and 50 per cent for those harbouring large predators. While estimating optimum stocking rates for such populations, about which no reliable estimates of natural mortality are available, it is felt that assumption of higher natural mortality rate would be desirable as a little overstocking would be less harmful than understocking. Fish Seed Committee of the Government of India recommended the stocking rate for reservoirs at about 500 fingerlings ha^{-1} of the size range 40 to 150 mm.

5

World Aquaculture Production

World aquaculture has grown from a few million tonnes annually in the 1950s to nearly 60 million tonnes in 2004. Statistics from FAO are used below to provide an overview of current world aquaculture production.

GROWTH OF AQUACULTURE

A noteworthy, but perhaps disturbing trend seen in the statistics for the Asian countries/territories from 1988 to 1997 is that the average aquaculture growth (APR) has been 10.4 per cent by weight and 9.0 per cent by value, but production has not shown steady growth, with year on year increases varying between 1.6 per cent in 1990 and 18.6 per cent in 1992. After peaking at 18-19 per cent in 1992-1993, the rate of increase has slowed down steadily to 5.7 per cent in 1997.

This observation coincides primarily with the trends seen for Chinese aquaculture. Some countries within the region have shown production decreases, with a steady decline of production through the period. These are Japan, China, Hong Kong8, Democratic Republic of Korea and Taiwan Province of China.

A striking trend in Japanese aquaculture has been the decline in production of the major species. For instance, the yields of Japanese amberjack have declined from a high of nearly 170 000 mt in 1995 to 147 000 mt in 1998, while laver dropped from a peak of 483 000 mt (1994) to under 400 000 mt (1998). Pacific cupped oyster production has also shown a steady decline. Of the major mariculture species, only silver seabream and Yesso scallop have shown trends towards rising production.

The annual rate of production growth has been consistently higher in China than in the rest of the region, reflected by the increased rate of difference in the compared production volumes. If aquaculture values are compared, the picture is slightly different. The difference in value is proportionately much less than that of production, confirming the effects of diversification and the maintenance of value in aquaculture in the rest of Asia. Indeed, in China the APR for value has grown and shown stability from 1990-1996, dropping off in

1997. In the rest of Asia, the value APR grew steadily from 1989, but has since declined regularly from 1994 onwards, even showing negative figures in 1996 and 1997. Theses circumstances meant that Chinese production alone had a total value higher than the rest of Asia from 1996.

DEMAND FOR AQUATIC PRODUCE

The projections for the average annual population growth rates in the region from 1998 to 2020 are:

- 0.8 per cent for East Asia, growing to 1.66 billion;
- 1.6 per cent for Southeast Asia, growing to 655 million;
- 1.7 per cent for South Asia, increasing to 1.72 billion; and
- Iran will have 89 million and Australia 22 million.

The total equals 4 124 million people or two-thirds of the earth's population. Just to maintain, region-wise, the supply of 17.2 kg of fish per capita (1995 average), the Asian Region would require more than 70 million mt of food fish in 2020, to be supplied from capture fisheries and aquaculture. A look at the situation by region would show differences, but all the same, the indications are for aquaculture to increasingly fill the fish supply requirements of the populations.

In China, aquaculture already contributes more than 60 per cent of the total fisheries production and, in 1995, the per capita supply of fish was 22.2 kg, implying a need for 32.2 million mt by 2020 assuming a constant per capita fish supply. If the same supply proportion were to be maintained, 19 million mt would be needed from aquaculture. Southeast Asia's average per capita supply is now 24.7 kg a year; with a population of 655 million it will require 16.2 million mt of food fish. In assuming that 30 per cent is to be contributed by aquaculture, almost 5 million mt of food fish need to be supplied by aquaculture. South Asia will need a total of 8.1 million mt of food fish. In 1997, only 2.33 million mt were produced by aquaculture, 97 per cent of this being food fish. If the same ratio is maintained, the South Asian Region will need 7.8 million mt by 2020.

With a total requirement of 35.7 million mt of cultured food fish forecast for 2020, the concern for supply is evident. Since the wild catch is stagnating and forecast to slow even further, aquaculture will have to increase its contribution beyond its current levels.

The Food and Agriculture Organization of the United Nations (FAO) predicts that global aquaculture is likely to show sustained growth in the medium term and could attain 35 to 40 million mt of finfish, crustaceans and molluscs by the year 2010. What this analysis clearly points out is that there is now a burden on aquaculture to meet the future demand for aquatic products, especially that the continued capacity to sustain further productivity increases in all the food production sectors is facing the following threats: losses in genetic resources and biodiversity (*e.g.* 4 per cent of fish and reptile species are threatened with extinction); loss of aquatic habitat such as coral reefs and mangroves, and severe depletion of commercial fish stocks (25 per cent of fish stocks for which data are available are either depleted or in danger of depletion, and another 44 per cent of fish stocks are being fished to their biological limit).

Aquatic resources are not unique in facing such degradation, as 9 million ha of land are extremely degraded, with their original biotic functions fully destroyed, and 10 per cent of the earth's surface is at least moderately degraded, and the availability of safe and clean water has declined steeply to 60 per cent (in 1996) of its 1970 value. These trends, which show no signs of changing, indicate that future production of food from aquaculture will face significant challenges in access to and use of resources.

Another crucial point that should be emphasized is that the countries in this region have the highest population densities in the world. There are many issues that aquaculture has to face, but the fundamental one for Asia is that of supply and demand. The critical concerns identified for resolution of this issue relate to social, economic, biological, environmental and institutional questions, which must be resolved with a minimum of conflict and most efficient use of resources. To put this into a positive light, aquaculture must be developed in a spirit of social harmony, environmental sustainability and economic progress.

TRENDS AND ISSUES

International Trade in Aquaculture Products

Information on the trade in aquaculture products is rarely separated from trade in fisheries products, making it difficult to analyse trends in the trade of aquaculture products. Comments here are therefore based mainly on fisheries products.

By volume, Asia is a net importer of fishery products, mainly because of Japan's trading patterns where an annual trade deficit (exports less imports) of more than 3 million mt has been registered since 1994. If Japan is excluded from the trade data, the developing countries of Asia would register a surplus of only 35 000 mt a year in the triennium up to 19979. The major net importers (with volumes over 100 000 mt) of fish are China, followed by the Philippines and Malaysia. The major exporters with volumes greater than 100 000 mt are India, Indonesia and Thailand. Of these, aquaculture production probably makes the biggest contribution in Thailand.

In terms of financial value, the countries with developing economies are responsible for more than half of the region's fishery exports, and there is a dramatic change in the pattern of imports and exports. Exports from developing countries in Asia are mainly high-value products that are directed primarily to developed economies. In the triennium ending in 1997, Thailand was the leading exporter in the region, exporting US$2.92 billion annually, a significant portion of which was cultured shrimp. Other major exporters with more than US$1 billion of fish products per year over the triennium include China, India and Indonesia. In terms of financial value, the countries with developing economies are responsible for more than half of the region's fishery export trade, which is directed primarily to developed economies. Between 1993 and 1996, Thailand was the leading world exporter of fish products, at annual values of around US$3.4 billion (although displaced by Norway's fishery exports in 1997). Japan is the leading importer of fish products in the world, with a value of US$15.5 billion in 1997.

United States absorbs about 10 per cent of the total value and these two countries and the European Community (EC) (including the value of the intra-EC trade) import 75 per cent (in value) of internationally traded fishery products.

Shrimp Trade

In financial value, shrimp is the most important traded aquaculture commodity in Asia, and Thailand is the world's main supplier of cultured shrimp, with an output of over 200 000 mt. In 1997, the shortage of shrimp on the world market was acute; but Asia, as a whole, was able to maintain its share of the United States market, and smaller exporting countries in the region such as Indonesia and China performed well, although China has of late rarely exported any shrimp.

Indian shrimp exports to Japan increased by 6.6 per cent to a record 59 100 mt, while total exports in the late 1990s have exceeded 100 000 mt, of which around half were cultured. The reasons for increased import to Japan included EU restrictions on Indian seafood, which started in August 1997.

In 1997, Japanese imports of shrimp fell by 7 per cent to 267 200 mt, the lowest figure in nine years, reflecting a downward trend that was apparent throughout 1997. The United States shrimp market was very strong, owing to the country's expanding economy and to the high value of the US dollar, and prices rose by 20 per cent in one year. United States imports expanded by 10 per cent in 1997, overtaking Japan as the world's major shrimp market for the first time. Within the region, for markets such as China (including Hong Kong), Singapore and Malaysia, the imports of fish and fishery products for domestic consumption continued to be lower in 1997 than in 1996 due to the economic problems experienced.

Seaweed Trade

Seaweed exports bring significant earnings to the Philippines, China, Indonesia and the Republic of Korea. The Philippines produced over 620 000 mt in 1997 of mostly Eucheuma seaweed, from which the colloid carageenan, which is the export product, is extracted.In East Asia, seaweed (Laminaria and Undaria) are also produced and particularly Undaria, exported.

Mollusc Trade

Mollusc exports from the region, including intra-regional trade, increased from 240 000 mt in 1988 to 335000 mt in 1997, peaking at 380 000 mt in 1996. Important exporting countries are China, Democratic Peoples Republic of Korea, Republic of Korea, Malaysia and Thailand. The predominant products are clams, scallops and oysters.

Finfish Trade

The trade in live marine fish, particularly reef fishes, has been growing steadily, spurring work on their aquaculture. Market analysis done in 1995 indicated that the total seafood market in Hong Kong and southern China, the main markets at present, was over 220 000 mt a year, within which the estimated annual demand for high-quality live reef fish was 1 600 to 1 700 mt. This study forecast compound growth rates of more than 12 per cent, in other words, doubling every six years.

In 1997, Hong Kong consumed 28 000 mt of live fish, of which groupers represent 35 per cent by weight and 50 per cent by value. Tilapia, filleted or chilled, is a growing export species. Taiwan Province of China supplied half of the 50 000 mt of tilapia consumed by the United States in 1998. Other exporters of tilapia to the United States market, but as yet in small quantities, are Indonesia and Thailand.

Singapore is also a significant market for high-value marine fish, although local cage farming has mainly satisfied its domestic consumption. To meet expected growth in demand, it has initiated the move to farm marine fish in offshore deep-water cages, targeting annual production of 40 000 mt that would be supplied with fry from local hatcheries.

The breeding and growing of ornamental fish and aquatic plants is also a growing industry. The value of world trade (wholesale and retail) in these commodities has been estimated at US$5 billion, generating employment and income to breeders and growers. Singapore, Malaysia and Thailand have well established and mature industries. Countries such as Sri Lanka and the Philippines have discovered this sector fairly recently, seeing that the activity offers opportunities for rural families to earn income from cultivating freshwater ornamental fishes.

QUALITY OF AQUACULTURE PRODUCTS

The importance of aquaculture production in Asia and in particular that of freshwater fish, most of which are sold in the domestic market, has a significant bearing on the concerns of human health associated with products from aquaculture.

The most common method of producing low-value freshwater fish in developing countries, especially in low-income food deficit countries (LIFDCs), is to use a combination of fertilizer application and supplementary feed. Polyculture of species that have complementary feeding habits is a general practice, especially with major and Chinese carps. However, the patterns of production are changing rapidly, with a large increase in more semi-intensive farming. There will be further emphasis on food safety issues in the future. In 1997, a WHO (World Health Organization)/FAO/NACA Study Group looked into the food safety issues associated with aquaculture and concluded that there were considerable needs for information, recognizing that such knowledge gaps could hinder the development process. Education is needed to increase the awareness of potential hazards in aquaculture and their management. The Study Group also recognized the difficulties in applying Hazard Analysis Critical Control Point (HACCP) procedures to small-scale farming systems and that food safety hazards associated with aquaculture vary by region, habitat and environmental conditions, as well as by methods of production and management.

In Asian shrimp production and processing, food safety is given increasing attention, largely because of the quality standards required by importing countries but also because of the higher profile given to environmentally friendly practices in the aquaculture sector.

AGREEMENTS REGULATING INTERNATIONAL FISH TRADE

International rules and regulations play a major role in governing the trade of fishery, affecting aquaculture products in both developed and developing economies. Several major international agreements and codes of conduct include sections that relate to aquatic animals and their products. In the Asia-Pacific Region, however, there is a wide difference in knowledge and expertise between different countries Furthermore, some countries in the region are members of organizations such as the Office International des Épizooties (OIE) and the World Trade Organization (WTO), and comply with the terms of such agreements, while others who are not signatories are obviously not bound by the terms of such agreements.

Regulatory and Management Frameworks for Aquaculture

Agencies Responsible for Aquaculture Development

In Asia, the primary responsibility for aquaculture production usually resides within one government ministry or department. By contrast, the management of the resources upon which aquaculture depends, particularly water and land, is normally the responsibility of other ministerial departments. This situation inevitably causes the potential for conflicts between different departments, and confusion or lack of clarity in the application of policy.

The increasing importance of aquaculture strongly argues for governments to give priority to developing clear, well-formulated, implementable and realistic national and local policies for aquaculture development, based on financial, social and environmental sustainability. As the private sector is the key to successful and sustainable aquaculture development, the views of industry should be taken into account in respect of policy formulation, research and development.

Participation

The adoption of a "participatory approach" is receiving increasing attention in both aquaculture planning and management. Experience with coastal fisheries management has shown, in general, that the failure to include coastal residents in natural resource management can lead to a lack of community compliance, resulting in resource depletion and conflicts.

This situation can be particularly acute when the government's capacity to enforce laws and regulations is limited.

A promising approach to this problem is "co-management", which involves the cooperation of the local community in both establishing and enforcing local management rules, with the requisite support from government. The co-management approach has proved useful for community-based management of some coastal capture-fishery resources, but the devolution of ownership and management of resources towards local people and communities should be explored for coastal aquaculture. The need to develop effective local arrangements for management of aquatic resources will also be stimulated further in some countries by the trend towards devolution of decision-making powers to local government bodies, as in the Philippines, Thailand and Viet Nam.

It is also apparent that the views of non-aquaculture groups need to be considered when formulating policy and implementing aquaculture development; without such views and contributions, some problems relating to conflicts of interest may be too difficult to fully resolve. Such issues are quite new, but there is evidence of increasing involvement of various aquaculture interest groups, including the private sector, local farmers and others, in aquaculture policy formulation. In some countries, the government and farmers have entered into discussion with NGOs and non-aquaculture groups for the formulation of management strategies. Examples have been seen in Sri Lanka, where NGOs have participated in aquaculture policy discussions, and in Malaysia, where input from various government, farming, academic and NGO interest groups was provided in the formulation of the shrimp and marine fish "Codes of Conduct".

Legislation

In many countries, aquaculture legislation is poorly developed and frequently consists of a few articles contained within a law regulating capture fisheries. During the last few years, interest has grown to develop a comprehensive regulatory framework for aquaculture whose goals would be to protect the industry, the environment, other resource users and the consumer, and to clarify rules for resource use. This consideration is driven by a variety of factors that include the following:

- Greater political attention to the sector as its economic importance and potential become more apparent;
- Realisation that inappropriate laws and institutional arrangements can severely constrain the sector's development;
- Evidence of environmental damage and social disruption as a result of the rapid and largely unregulated expansion of shrimp farming; and
- A growing emphasis on examining production methods as a means of improving the quality and safety of aquaculture products.

While capture fisheries are generally regulated by a single government department or ministry, aquaculture is frequently regulated by many agencies under a variety of laws. In this circumstance, the formulation of a comprehensive regulatory framework for aquaculture is legally and institutionally complex. Typically, the tasks involve drafting or amending legislation that addresses a variety of issues such as land use planning and tenure, water quality and environmental issues, fish movement and disease, pharmaceutical and chemical use, food quality and public health.

It also requires clarification of responsible authorities and the establishment of the arrangements to ensure cooperation and coordination between many different institutions that have jurisdiction over the item or issue concerned.

INTERNATIONAL DEVELOPMENTS

The adoption of the Code of Conduct for Responsible Fisheries (CCRF) by the 1995 FAO Conference promises to provide a significant influence on the development of regulatory systems for aquaculture in the coming years. Article 9 of the Code deals with aquaculture development and sets out a wide range of relevant principles and criteria. In addition, the "Jakarta Mandate", which was adopted by the second Conference of the Parties to the Convention on Biological Diversity in 1995, provides useful guidance regarding aquatic biodiversity and related environmental aspects that should be taken into account when developing coastal aquaculture.

During 1994 and 1995, several regional workshops10 were organized to examine environmental, legal and institutional issues associated with aquaculture. Recently, FAO published a compendium on aquaculture and inland fisheries legislation in the Asian Region. Although new national laws to regulate aquaculture comprehensively may be desirable in many countries in the region, other options are now being explored. This is because the time required to develop and pass new comprehensive legislation is likely to be several years, while the rapid development of the sector has created an urgent need for regulation. These options include the enactment of regulations that exist under existing legislation and the application of voluntary approaches such as guidelines and codes of practice. Creating incentives to encourage compliance or disincentives for noncompliance or failure to join the scheme can enhance the effectiveness of voluntary codes of practice. One example is the policy that was adopted in India in 1995 by the National Bank for Agriculture and Rural Development (NABARD), which was a major source of refinancing for brackishwater prawn farming in the states of Andhra Pradesh and Tamil Nadu. This policy sets out the conditions to be met before NABARD will provide credit to banks for aquaculture developments. Similarly, Malaysia is considering the requirement that aquaculture enterprises which are given preferential "pioneer" tax status must adhere to a code in order to retain the various tax incentives accorded. Clearly, there is scope for increasing the use of such conditional public and private-sector funding to encourage the application of sustainable aquaculture practices.

The Australian State of Tasmania has recently passed legislation that provides another interesting example of what can be done to address this issue. The new laws (notably the Marine Farming Planning Act [1995] and the Living Marine Resources Act [1995]) impose that the preparation of marine farming development plans must cover areas, rather than sites, and provide for broad community participation in the preparation of these plans. An environmental impact assessment must also be done, with the establishment of a marine farming zone, before site leases are granted to marine farms.

Malaysia is developing an integrated regulatory system for aquaculture, in the short to medium term, but without formulating a specific aquaculture act. Instead, new regulations will be incorporated under the existing Fisheries Act, and a voluntary Code of Responsible Aquaculture Practices will be introduced for cage culture and shrimp farming. These measures will be supported by incentives, and institutional structures will be strengthened to ensure the effective monitoring and continuing formulation of aquaculture policy. Many regulatory systems could be improved by including mechanisms that are designed not only to prevent or reduce the risk of environmental harm but also to help right any damage that may occur. The Tamil Nadu Aquaculture (Regulation) Act of 1995 not only set out conditions to improve the siting and management of aquaculture facilities, but also established an "Eco-Restoration Fund", funded partially by deposits from aquaculturists, to be used for remedying any environmental damage caused by aquaculture farms. In addition, the constitution of the Aquaculture Authority of India has been made for the regulation of shrimp aquaculture.

PRODUCTION BY ENVIRONMENTS AND SPECIES CATEGORIES

Aquaculture is considered by FAO to be done in three environments: freshwater, brack-ishwater, and marine. In 2004, total aquacultural production for the three environments were 25.8 tonnes in freshwater, 3.4 tonnes in brackishwater, and 30.2 million tonnes in marine environments. The data are summarized by major species groups. Freshwater aquaculture is largely fish culture, crustacean culture dominates brackishwater aquaculture, and aquatic plants (primarily seaweeds) and mollusks are the main marine aquaculture crops. If only aquatic animals are considered, aquaculture production in freshwater is considerably greater than that of brackishwater and marine environments combined. Further analysis of the FAO data reveals that carp species dominate freshwater production, but there also is significant production of unspecified fish species and tilapia. Production of Pacific white shrimp (*Litopenaeus vannamei*) and black tiger shrimp (*Penaeus monodon*) dominates brackishwater aquaculture. Bivalve mollusks are the main categories of marine animal culture, but Atlantic salmon (*Salmo salar*) is also significant. Of course, marine plant production is slightly greater than bivalve mollusk production in marine aquaculture.

Geography of Aquaculture

The world is divided into six aquaculture regions by FAO. The following are the regions and the 2004 production of each: Oceania, 139,273 tonnes; Africa, 570,113 tonnes; North America, 955,178 tonnes; South America, 1,137,825 tonnes; Europe, 2,238,683 tonnes; Asia, 54,367,372 tonnes. Asia provides 91.5 per cent of world aquaculture production.

Although many nations have aquaculture, relatively few nations provide most of the world's production of cultured aquatic organisms. Nine of the top ten and twelve of the top 20 aquaculture nations are in Asia. According to FAO statistics, China is responsible for more than 50 per cent of world aquaculture production, with more than 41 million tonnes annual production. The FAO can officially report only fisheries data provided by governmental sources in each nation. Most experts agree that the Chinese government has traditionally inflated their aquaculture data—possibly by a factor of 2 or 3. Nevertheless, even if the Chinese data are inflated fivefold, China still has more than twice the aquaculture production of India, the second largest aquacul-ture producer.

Aquaculture Products in International Trade

The top ten species and species groups produced by aquaculture include salmonids, tilapia, shrimp, oysters, clams, scallops, and mussels, popular with consumers in developed nations where a high percentage of seafood is imported. For example, about 2.1 million tonnes of seafood are consumed annually in the United States, of which about 80 per cent is imported. Average per capita consumption of seafood in the United States is 7.4 kg, and 10 species or species groups account for over 90 per cent of the total consumption. Aquaculture is an important source of shrimp, salmon, catfish, crab, tilapia, clams, and scallops for United States consumers. Channel catfish (*Ictalurus punctatus*), is almost totally from domestic aquaculture. Two species, red swamp crayfish (*Procambarus clarkii*), known in the United States as crawfish, and rainbow trout (*Oncorhynchus mykiss*), also are popular with United States consumers, and these two species are produced mainly in domestic aquaculture.

CONTRASTS BETWEEN UNITED STATES AND INTERNATIONAL AQUACULTURE

There are, however, major differences between aquaculture in the United States and other developed countries and aquaculture in developing nations. In developed nations, most seafood is imported from developing countries, and aquaculture is limited mainly to a few species of moderate or high value for domestic markets. For example, channel catfish, rainbow trout, and crawfish produced in aquaculture in the United States are almost exclusively marketed domestically. Several countries in the European Union produce trout, salmon, and bivalve shellfish by aquaculture. Although some of this production is exported, much is for the European market. Carp production in Europe is for domestic markets. In developing nations with large aquaculture sectors, there is greater variety of culture species than in developed nations. Although most of the production is consumed domestically, some developing countries export large quantities of aquacultured shrimp, tilapia, salmon, and several other species.

Farms in developed countries usually must be fairly large to be profitable, and there are relatively few producers. For example, in 2006 there were 1,035 catfish farms in the United States and the average farm size was 67 ha. Labour is expensive, so mechanization and automation is used as much as possible. Even family farms must invest in the latest technology to be profitable. There are tens of thousands of small aquaculture farms in many developing nations. In pond culture, farms typically have water surface areas of 0.25 to 5 ha. Other production systems also are much smaller than those in developed nations. There are some extremely large shrimp, tilapia, and salmon farms in developing nations, but the combined production by small farms in many nations, especially in Asia, greatly exceeds that of large farms.

Small-scale farmers tend to be poor and live in rural areas. They are poorly educated, lack knowledge about aquacultural technology, and are unaware of environmental issues. Production practices often are substandard, and few producers use BMPs to lessen negative environmental impacts. Large farms usually hire highly qualified managers and use better production practices than small farms. Because labour is inexpensive in developing countries, large farms are not as mechanized as those in developed countries. Nearly all large farms in developing countries focus on the export market. There often are dense concentrations of small aquaculture farms in developing nations. The environmental implications of such dense clusters of farms are obvious, and the likelihood of negative impacts often is exacerbated by lack of effective governmental regulation of effluents, wetlands, predator control, water use, and other factors. The combined effect of many small farms on the environment can be as great or greater than that of a large farm.

Aquaculture facilities in developed nations also tend to be concentrated in specific areas. In the United States nearly all catfish farms are located in a few counties of Alabama, Arkansas, Louisiana, and Mississippi. In Alabama about 50 per cent of catfish farms in the state are on the catchment of a single creek, but the farms occupy only 7.5 per cent of the surface area of the catchment. The largest portion of the United States trout industry is located along a stretch of the Snake River in Idaho. Nevertheless, the facilities usually are separated by distances of 1 to several km. Environmental regulations also are imposed on aquaculture in wealthy nations, and there are few alleged or documented negative environmental impacts of aquaculture.

Developed nations generally have more equitable social conditions than found in developing countries. In many poor countries, agricultural workers may not be paid a legal minimum wage, working and living conditions on farms may be inadequate, and there may be no benefits or job security. Large aquaculture companies may be given priority over other stakeholders for use of land, water, and other resources, resulting in conflicts with local communities. Companies may hire laborers from other places in preference to local workers. Child labour also may be used at some farms. Small-scale producers usually do not hire workers, but underaged family members are expected to work on farms. This is acceptable provided children are not denied the opportunity for an education. Obviously, social injustices are more likely to result from aquaculture in developing nations than in the United States or other developed nations.

Environmental NGOs continue to criticize aquaculture on environmental and social grounds, but discussions with officials in several of these organizations suggest a change in attitude. They have come to realize that aquaculture is a necessary component of the world's food production system, and that when done properly, aquaculture can be more environmentally responsible than fishing. Fishing is analogous to hunting and gathering, while aquaculture is a form of agriculture. Humans no longer depend on hunting and gathering as a primary source of food and fibre from terrestrial sources. Likewise, we should increasingly turn to responsible aquaculture as an alternative to fishing for sources of aquatic food.

CODES OF CONDUCT AND BMP PROGRAMMES

There are a number of initiatives to improve the environmental and social performance of international aquaculture. These range from programmes encouraging voluntary adoption of BMPs to certification programmes in which

BMPs are implemented as a means of complying with verifiable environmental and social standards. There has been widespread use of BMPs to improve resource use and lessen the amount of non-point source pollution in traditional agriculture. Thus, the initial response of promoting codes of conduct and BMPs in response to the criticism of environmentalists over negative impacts of aquaculture was not surprising. The efforts mostly were initiatives of aquaculture associations, government fisheries and aquaculture agencies, and international organizations that support aquaculture development. It is noteworthy that the promotion of better practices began well before there was concern in the international market about environmental and social issues related to aquaculture.

FAO Involvement

The FAO has a long history of promoting better aquaculture practices to increase production, improve efficiency of resource use, and avoid negative environmental and social impacts. The FAO Fisheries Department published a *Code of Conduct for Responsible Fisheries* that has served as guiding principles for many of the recent responsible aquaculture programmes. Shrimp aquaculture has been a particular focus of FAO. Expert consultancies were held in Bangkok, Thailand; Rome, Italy; and Sydney, Australia to discuss the environmental impacts of shrimp farming and to develop best management practices for preventing these impacts.

The Consortium on Shrimp Farming and the Environment

The World Wildlife Fund (WWF), Network of Aquaculture Centres in Asia-Pacific (NACA), and World Bank joined with the FAO and the United Nations Environment Programme (UNEP) to form a consortium to investigate shrimp farming practices worldwide. The consortium conducted case studies of shrimp farming in several countries. For example, in Ecuador, studies were conducted on capture of wild postlarvae, farm effluent composition, and effects of shrimp farming on mangroves. Thematic studies that cut across countries were prepared for topics such as chemical use, inland shrimp farming, shrimp diseases, etc. Many of the case studies have been posted at the NACA web site. The findings of the case studies were used to develop better management practices for shrimp farming.

Government Programmes

Aquaculture products are important sources of foreign exchange in many countries. Thus, governments of nations with significant aquaculture have the dual task of protecting the environment and promoting the export of aquaculture products. Some countries have programmes to assist producers in developing and adopting codes of conduct and BMPs for responsible aquaculture. The government of Madagascar developed a code of conduct for shrimp farming with recommended practices, including some legal aspects to be considered later. The Thailand Department of Fisheries devoted much effort to promoting a code of conduct (CoC) programme for marine shrimp farming. Recently, the Thailand Department of Fisheries also initiated a good aquaculture practices (GAP) programme for Thai shrimp farming that is less strict than the CoC programme. Producers can choose between the CoC and GAP programmes (Waraporn Prompoj, Thailand Department of Fisheries, personal communication). The Malaysia Department of Fisheries also made a CoC for shrimp farming that is an example of general guidelines for voluntary adoption by producers.

Industry Efforts

Aquaculture industry associations in several nations have developed environmental policies, codes of conduct, or codes of practices for their members. One of the first and best-known programmes is the code of practices of the Irish Salmon Growers Association. Some other examples are programmes of the British Trout Association, Ornamental Fish Industry, Washington Fish Growers Association, Aquaculture Foundation of India, Pacific Coast Shellfish Growers Association, and the Association of Scottish Shellfish Growers. The Australian Prawn Farmers Association developed one of the first codes of practice for shrimp culture.

The Global Aquaculture Alliance and several other organizations have since developed codes of practice for responsible aquaculture and especially for responsible shrimp farming.

Voluntary Adoption of BMPs

Because of the huge amount of attention given to the negative environmental impacts of aquaculture and promotion of better practices, there no doubt has been voluntary adoption of BMPs by individual farms. It is unlikely that voluntary adoption has been widespread or systematic, because BMP implementation depends on the individual farmer's perception of the effect of adoption on farm profitability.

It is usually easier to assess the costs of BMP implementation than it is to measure benefits, so farmers often have a distorted picture of the farm-level economics associated with BMP adoption. If they perceive that BMP adoption is too costly, they will not change their behaviour. BMPs that are directly associated with better production efficiency (such as practices that improve feed conversion) will be more readily adopted than practices that are management-intensive or require investment in additional infrastructure. No studies have been conducted to determine the degree of independent, voluntary adoption of aquaculture BMPs.

RETAILER PURCHASING POLICIES

Several large retailers have developed purchasing policies and production standards for use in sourcing responsible aquaculture products. Examples are Marks and Spencer and Tesco in the United Kingdom, Carrefour in France, and Wegmans Food Markets, Inc., and Bon Appétit Management Company in the United States. Retailer programmes tend to focus on the production process, product specifications, and product quality. However, they generally include environmental and social requirements. Standards may vary considerably among different companies. Standards may be agreed upon between buyer and producer and not disclosed to the public. Some retailers, *e.g.*, Wegmans and Bon Appétit Management Company, apparently will disclose their policy and standards. Retailers have their own specialists to audit producers or they rely on auditing companies. Retailer programmes can be important in providing safe, high-quality products and in reducing negative environmental and social impacts. The standards, however, are developed specifically for the retailers' marketing objectives and customer base. The standards may not be revealed to the public, and development of standards is not always a transparent process. Some retailers are hiring professionals with experience related to aquaculture and environmental issues. Others are contracting with environmental NGOs to obtain assistance in developing purchasing policies and standards. Of course, this approach possibly also is a means of obtaining an endorsement from the NGO.

The primary purpose of retailer programmes is for a company to differentiate itself by developing its own standards. This suggests that the retailer understands and controls the suppliers. The standards also suggest that the retailer is seeking a product specifically for its customers. These programmes are marketing strategies intended to satisfy a company's customer base and to demonstrate that the company is concerned about the environment and social issues.

Government Programmes

The French Label Rouge programme is an example of government-controlled certification. This programme certifies many food products for the French market and it has standards for aquaculture certification. It is run by an association of industry and certifying organizations with oversight by the French government. Like retailer programmes, it focuses on product quality but it has environmental and social standards.

In Thailand, the Department of Fisheries has developed two BMP programmes for shrimp culture that have been promoted (so far unsuccessfully) as a means of assuring buyers that shrimp have been produced responsibly. One is called the *Thailand Marine Shrimp Farming Code of Conduct (CoC)* and the other is known as the *Good Aquaculture Practices (GAP)* for marine shrimp farming. Both programmes require implementation and verification of specific BMPs but do not have clearly defined standards. Based on the practices required for compliance, the CoC programme appears more rigorous than the GAP. The Department of Fisheries has trained inspectors for these programmes and many farms have been certified. Representatives of the shrimp culture industry were involved in discussions related to formation of both programmes, but apparently few other stakeholders had input. The method for inspecting farms for compliance with certification requirements does not appear to be a true third-party system.

The CoC and GAP programmes have created an interesting situation. The DOF was encouraged by the World Bank and other outside groups to develop the CoC and GAP as a proactive approach. The Thai Department of Fisheries wants its certification programme to be accepted by foreign shrimp buyers, and it desires sole responsibility for certification of farms in Thailand.

So far, foreign shrimp buyers are not willing to accept the Department of Fisheries programme as a credible means of sourcing environmentally responsible shrimp. Some in the Department of Fisheries feel slighted because of the lack of confidence in the CoC and GAP, and they have responded by not cooperating with those interested in installing other certification programmes.

The FAO has announced an initiative to develop international guidelines for aquaculture certification. The first expert consultancy was held in Bangkok, Thailand, from 27–30 March 2007, and at least three more consultancies are planned for other venues. This effort possibly will provide suggestions for inclusion of national interest in development of standards and administration of certification programmes.

The shrimp farming code of conduct developed in Madagascar can be considered a type of government certification. Many of the practices presented in this document are legally binding. This code prohibits shrimp farms in mangrove areas, specifies minimum distances between shrimp farms, prohibits imports of broodstock and postlarvae, places limits on production intensity, and requires effluent monitoring. Adoption of the practices by all producers has allowed shrimp to be sold under a "Product of Madagascar" label and receive a premium price.

An Industry Certification Programme

The Global Aquaculture Alliance (GAA) developed certification standards for shrimp aquaculture and licensed these standards to the Aquaculture Certification Council (ACC). The GAA has shrimp certification standards for hatcheries, farms, and processing plants. The ACC has held several training programmes to train inspectors, and many farms have been certified. This programme was given a great boost recently when Wal-Mart signed an agreement with ACC to sell only ACC-certified shrimp as soon as the programme can be widely implemented (Chamberlain 2005, 2006). The GAA currently is developing standards for certification of several fish species at the request of Wal-Mart and several other buyers.

The relationship between GAA and ACC has been the subject of considerable controversy. The GAA was established in 1997 with the stated mission "to further environmentally responsible aquaculture." The GAA developed BMPs for voluntary adoption by shrimp farmers. This effort led to the idea of certification. Certification standards were developed for which producers would have to install BMPs to achieve compliance. This programme was called *Best Aquaculture Practices (BAP)* certification. The programme has been widely criticized by NGOs for not involving a wide range of stakeholders in developing the standards.

This criticism is not entirely valid, for several environmentalists were invited to review the standards, but only a few did. The ACC was developed by GAA as the certifying body for the BAP programme. Environmental NGOs complained that GAA and ACC were in reality the same organization because there was much overlap on the board of directors of the two organizations. The two organizations have gradually become independent bodies, and today, only one individual is a member of the board of directors of both organizations. The ACC and GAA have distanced themselves and operate independently, and the criticism, in my opinion, is no longer valid.

The ACC developed the procedures for verification of compliance with BAP standards and trained inspectors for the programme. The environmental NGOs correctly argue that the ACC is not a true third-party certification programme because the inspectors are not from an independent organization. At Wal-Mart's request, Conservation International (CI) evaluated the GAA standards and the ACC procedures. GAA and ACC incorporated changes suggested by CI, but the ACC certification programme still is viewed with suspicion by most environmental NGOs. Originally, certification was promoted as a way of obtaining a higher price for aquacul-ture products. However, Wal-Mart apparently does not plan to pay a higher price for ACC-certified shrimp. The producer will have to bear the extra cost of certification to assure sales to Wal-Mart. Environmentalists would like to see producers benefit from a higher price for their products. Nevertheless, they are probably willing to accept the Wal-Mart model because it will result in a much wider need for certification than would result from high-end markets for certified aquaculture products. Wal-Mart's competitors already are initiating efforts to market certified aquaculture products. It is not yet clear as to whether they will pay producers more for a certified product.

The WWF conducted a study of the issues that should be considered in developing certi-fication standards for shrimp, salmon, catfish, tilapia, trout, oysters, mussels, clams, scallops, abalone, and seaweed. Reports have been prepared on the use of fishmeal in aquaculture and on the markets for certified aquaculture products. Stakeholder dialogues have been conducted for salmon, channel catfish, tilapia, and shellfish certification issues, and additional dialogues are scheduled. The WWF received additional funding for this effort in early 2007, which has allowed them to hire several specialists to manage the dialogues for different species or species groups. The timeline for the project calls for stakeholders to reach agreement on issues and develop the initial certification standards for the different aquaculture species by the end of 2009.

The WWF programme emphasizes adoption of better practices to improve the performance of aquaculture. Clay also suggests use of the term *better management practices* instead of *best management practices* because practices are evolving. Adoption of better practices will allow farms seeking certification to comply with a system of numerically verifiable standards.

Although the WWF is devoting much effort to aquaculture certification, it will not be the certifying body. A suitable organization to oversee the certification programme will be found and the WWF effort entrusted to it. This is the same way that the WWF and its donors operated in developing the Marine Stewardship Council (MSC) and the Forest Stewardship Council (FSC). The MSC develops standards for sustainable fisheries, but an independent certifying body is responsible for determining whether a fishery complies with the standard. This is true third-party certification. The FSC operates in much the same way as the MSC. There have been discussions among the GAA, ACC, WWF, and MSC about possibilities for combining efforts. A certification programme developed and operated through collaboration among the WWF, GAA, and ACC would probably be more widely acceptable than either an NGO or industry programme alone. However, the author is of the opinion that aquaculture certification and fisheries certification should not be combined. Aquaculture is potentially more responsible than fishing, and the aquaculture industry should promote their products separately.

Certification

The purpose of aquaculture certification is to provide products certified to have been produced by environmentally and socially responsible methods. There are several generally accepted requirements for developing certification programmes. The standards should be developed through a broad, deliberate, and transparent process that involves all stakeholders. These standards should be available to the public. Producers participating in certification must comply with the standards, and compliance must be verified through regular and unannounced inspections by a qualified third party. Non-compliance with certification requirements must be corrected within a specified period, or certified status is revoked. Product traceability is a key feature of certification because it must be possible to trace a certified product back to the farm of origin.

ISO Certification

Certification by the International Standards Organization (ISO) has been achieved by several aquafeed plants and processing plants for aquaculture species. A few shrimp and fish farms also have obtained ISO certification. There are several types of ISO certification; the applicable ISO programmes for the aquaculture industry are 9001 (quality), 14001 (environment), and 22000 (food safety).

ISO currently does not recommend aquaculture production standards, but the programme provides a procedure through which producers can develop and implement practices for compliance with standards from another source. In ISO certification, an organization says what it will do and then proves it does what it says it will do. Retailers and other buyers of aquaculture products likely view ISO certification favorably because it suggests that farms and processing plants are well managed. ISO certification is not proof of environmental and social responsibility, because the entity undergoing certification often makes its own standards without stakeholder involvement. Nevertheless, ISO certification is established and respected, and most environmental NGOs probably have a positive opinion of it.

Organic Certification

Organic certification of terrestrial food products has a long tradition. The primary purpose of standards is to verify that products are organically produced, but a few environmental and social standards usually are included. Organic certification recently has been extended to aquaculture products by several organizations.

The most popular ones are Naturland (Germany) and Soil Association (United Kingdom). The International Federation of Organic Agriculture Movements (IFOAM) developed aquaculture standards for inclusion in the IFOAM Basic Standards in 2005. The French Ministry of Agriculture and Fisheries is working on organic standards for aquaculture products, but presently organic aquaculture standards for France are based on agricultural standards. For example, OSO in Madagascar and the French auditing company Bureau Veritas developed organic shrimp standards that were validated by the French Ministry of Agriculture and Fisheries.

The United States Department of Agriculture (USDA) is developing standards for organic aquaculture under the National Organic Standard Board (NOSB). The Aquatic Animal Task Force has developed draft organic standards that have been approved by the NOSB and will now undergo interagency review. Once standards are approved, the USDA National Organic Programme will publish an organic aquaculture certification rule in the Federal Register. After implementation of the rule, all organic aquaculture products sold in the United States must comply with USDA organic standards.

EurepGAP Certification

The EurepGAP programme is promoted as responding to consumer concerns on food safety; animal welfare; environmental protection; and worker health, safety, and welfare. This programme, administrated from offices in Germany, was developed initially for traditional agricultural products. EurepGAP recently extended its standards to include aquaculture.

EurepGAP has an *Integrated Aquaculture Assurance* document with which all certified facilities must comply. This document does not have clearly stated standards. Instead it has several general principles related to worker safety, facilities management, chemical use, animal welfare, environmental protection, product quality, and product testing.

It requires the preparation of many action plans and has a huge record-keeping component. A total of 213 control points are identified in the basic aquaculture assurance document, and each must be audited for compliance with stated criteria. Some control points have multiple criteria. For each individual aquaculture species or species group, there is an add-on module for specific control points.

The add-on module for shrimp has an additional 32 control points. There are major, minor, and recommended control points. A facility must comply with all major criteria and 90 per cent of minor criteria, but compliance with recommended criteria is not mandatory. Although the EurepGAP programme is extremely detailed, it is much weaker on environmental and social requirements than most other programmes. The requirements for written plans and record keeping are complicated. Only a large farm capable of dedicating one or more employees to the tasks could comply with the criteria. The author has not heard environmentalists comment on the EurepGAP programme, but it seems unlikely that they would find this programme satisfactory for lack of social standards. Moreover, it is unlikely that producers would choose such a complex programme over a simpler one unless forced to do so by buyers. EurepGAP was a presenter at the 27–30 March 2007 FAO consultancy on international certification standards for aquaculture in Bangkok, Thailand. It appears that EurepGAP will develop specific aquaculture certification programmes for different countries. For example, a presenter from China talked about a EurepGAP-China aquaculture certification programme "based on the realities of Chinese aquaculture." This suggests that EurepGAP is planning to make changes in its programme that will facilitate certification of small-scale producers in developing countries.

CONSTRAINTS IN RESPONSIBLE AQUACULTURE PROGRAMMES

The responsible aquaculture movement is based on the premise that consumers in developing nations desire fish, shrimp, and other aquatic food products from environmentally and socially responsible sources. Therefore, retailers, restaurateurs, and other larger buyers are seeking such products. Environmental NGOs are interested in promoting environmental and social stewardship. Thus, they see aquaculture labelling programmes as an effective way of promoting their policy agenda. International development agencies such as the World Bank, Asia Development Bank, and FAO are committed to environmental stewardship and programmes by buyers to source responsible products that are compatible with that goal. Processors and exporters of aquaculture products would obviously participate in responsible aquaculture as a business opportunity. However, despite the popularity of responsible aquaculture, there remains perplexing issues that are seldom considered.

Extending BMP and Labelling Programmes to Small-Scale Producers

Most aquaculture is conducted in developing countries, and a large amount of the production is from small farms. Although much of the production of small farms is consumed by farm families or sold in the local market, some of it enters the export market. The large demand for imports of aquaculture products in wealthy nations cannot be satisfied by the production of large farms alone. However, the current effort on aquaculture labelling is directed almost exclusively at large farms. The cost of implementation of practices, analyses, and record keeping necessary for compliance with certification or other labelling programmes and the expense of third-party inspection is more than most small-scale producers can bear.

In 2006, the GAA convened an expert consultation in Bangkok, Thailand, on issues related to certification of small shrimp farms in Asia. The participants agreed that small farms need to be arranged in clusters to reduce the costs of compliance and verification. Moreover, it will require much work to develop procedures for establishing and administrating these clusters. The situation will differ from country to country and a single model for cluster farms would be unworkable.

The participants also concurred that standards for ACC certification would require modification for application to small farms. Small-scale producers would often not have title to their land, few would be able to demonstrate social stewardship, and effluent monitoring and other analyses would be excessively expensive. Small-scale farmers may integrate aquaculture with poultry, swine, or cattle production. Water supplies for small-scale farms may be highly polluted, and farm families may reside beside ponds and discharge domestic waste into them. These scenarios raise food safety issues that must be addressed by labelling programmes. The GAA has charged a committee with developing certification standards for cluster farms. The objective is to require the same degree of environmental and social stewardship for small farms as expected for large ones, but this may prove an impossible task.

Some feel that environmental concerns are being favored over social issues in BMP and labelling programmes. In a policy analysis of the ongoing effort to apply BMPs to shrimp farming, Béné suggests that the BMP effort is a scientific approach that, by basically ignoring social issues, has allowed a number of key stakeholders to refocus the debate on scientific solutions, thereby preventing other groups concerned with more intractable social and political issues from engaging in the policy process. Béné apparently feels that widespread use of labelling programmes will further marginalize small-scale producers. Although Béné does not offer any solutions, he hints that policy making in shrimp farming is too political to be reduced to the application of BMPs and other technical interventions. Béné's eloquent argument could be applied equally well to other forms of aquaculture, but it is not clear how to avoid some of his concerns and yet progress towards an environmentally responsible aquaculture. New also expresses concern that the responsible aquaculture movement will further marginalize small-scale producers. He stresses that responsible aquaculture should be profitable but also have a conscience.

Demonstration of Benefits

Programmes promoting voluntary adoption of BMPs and labelling programmes requiring verifi-cation of BMPs or compliance with standards are in their infancy. Studies of the adoption of these programmes have not been made, but it is certain that current participation represents only a minuscule proportion of aquaculture production. Most consumers in the United States presently cannot find environmentally labeled aquaculture products without shopping around, because few supermarkets and restaurants carry them. Moreover, there is scant documentation that adoption of BMPs and compliance with labelling programmes has environmental and social benefits.

Verification of environmental and social benefits would require carefully designed studies. In some places, comparisons could be made between farms participating in BMP and labelling programmes and those not participating. These comparisons could include variables such as efficiency of water use, fishmeal and protein use in feeds, feed conversion rate, effluent volume and quality, wetland destruction, and social issues such as compensation and working conditions for workers and relationship with local communities. In other places, it may be possible to make such comparisons before and after implementation of BMP or labelling programmes at a facility. However, aquaculture often is only one of several activities that can influence environmental and social conditions. In such situations, negative impacts caused by other resource users could mask positive effects that might result from better aquaculture practices.

Constraints in assessing the benefits of implementation of BMPs and compliance with labelling programme standards could be overcome by use of indicators (Boyd 2005a, b; Boyd 2006; Boyd and Polioudakis 2006). Effectiveness of voluntary adoption of BMPs could be revealed by improvements in values of indicators. Indicators could reveal ranges in the use of water, land, protein, fishmeal, energy, and other resources and in the amount of wastes produced per tonne of production for different species and culture methods. Evaluation of these ranges could be used to establish minimum acceptable values of each indicator for compliance with labelling programmes. Of course, social issues and some environmental concerns such as facility siting, predator control, exotic species, antibiotic and chemical use, and general sanitation possibly cannot be addressed by indicators.

Discrepancies in Programmes

Differences among sites, culture systems, production methods, and species also raise issues about aquaculture labelling. With regard to sites, mangroves and other wetlands are prohibited in responsible aquaculture programmes as sites for new farms or expansions. However, many existing farms were constructed in wetlands. Such farms obviously should be encouraged to adopt better practices. However, should they be allowed to participate in labelling programmes?

The potential for water pollution varies among culture systems and increases in the following order: ponds < flow-through systems < cages and net pens. Practices are available for reducing the pollution load in effluents from ponds and flow-through systems, but nothing can be done to reduce nutrient and organic matter release from cages other than to lessen feed input. At some sites, dilution and dispersal of waste may be adequate to avoid water pollution by cage and net-pen culture. Thus, should cage and net-pen facilities be denied access to labelling programmes or should those that are sited well be allowed to participate?

Water exchange has been used to flush ponds and improve water quality. This practice shortens hydraulic retention time and lessens effectiveness of natural processes in assimilating wastes. Should water exchange be strictly prohibited, restricted, allowed for emergencies, or allowed on a case-by-case basis in labelling programmes? The culture of some species is more likely to have serious negative environmental impacts than the production of other species. Boyd *et al.* (2005) suggested that contention with issues related to certification of some common species or species groups would increase as follows: seaweeds < bivalve shellfish < abalone < channel catfish < tilapia < trout < shrimp < salmon.

As there are differences with respect to certification issues among sites, systems, methods, and species, it is not reasonable to consider all certified aquaculture products equal with regard to environmental responsibility. For example, certified tilapia from ponds might have a better environmental record than certified tilapia from cages. The environmental impact per unit of production might be greater for salmon cultured in cages at a certified facility than for channel catfish grown in ponds at a non-certified farm.

The problems of organizing small-scale farms for certification and harmonizing cer-tification requirements for large- and small-scale facilities has already been mentioned. However, when compared with discrepancies related to sites, systems, methods, and species, differential treatment of small-scale producers in labelling programmes would appear to be a minor issue.

The fact remains that most aquaculture in the developing world will not be included in the responsible aquaculture movement because programmes are directed at facilities producing for the export market. National governments should take the responsibility for regulating aquaculture facilities to minimize negative environmental impacts. Moreover, social issues should also be addressed by national governments. Unfortunately, in many nations with a large aquaculture sector, national governments have neither the resources nor the will to deal effectively with environmental and social issues. Thus, the benefits of programmes for responsible aquaculture likely will be eclipsed by the negative impacts of domestic aquaculture.

Influence on United States Producers

Most foreign producers have lower production costs than producers in the United States because they have access to cheaper labour and are subject to fewer regulations. Aquaculture imports may compete directly with the products of United States aquaculture. For example, basa *(Pangasius bocourti*) and tra (*Pangasius hypophthalmus*) from Vietnam compete with channel catfish. Catfish farmers succeeded in preventing basa and tra from being sold in the United States as catfish and an antidumping tariff also was imposed. However, China has begun to ship channel catfish to the United States and governmental intervention into this trade is unlikely.

Widespread adoption of labelling programmes by retailers, restaurateurs, and other buyers would require both domestic and foreign producers to comply with requirements of these programmes. As a general principle, participants in labelling programmes must exceed minimum legal requirements for environmental and social stewardship. Aquaculture in the United States is regulated more strictly than in developing countries. It should be much easier and cheaper for a United States producer to comply with labelling programme requirements than it would for a producer in a country with few regulations. Thus, labelling should make United States aquaculture more competitive.

6

Major Physical Resources Required in Aquaculture

The major physical resources required in aquaculture are energy, land, water, and feed. All are finite and the impacts of aquaculture on resource availability depend on the overall rate and efficiency of use. Aquaculture may also impact resource availability by altering, rather than using, a resource. For example, flow-through aquaculture systems consume almost no water but they alter its quality by adding wastes produced during culture. Effects of resource alteration must be managed so aquaculture does not diminish the value of the resource for some other use.

Resources are also subject to conflicting demands on their use and, ultimately, these conflicts are difficult to address because they involve trade-offs. The fundamental question is whether the use of a particular resource in aquaculture has more social or economic value than its use in some other activity. The conflicting or alternative use might be in another food-producing sector of agriculture or it might be an environmental function, such as the biodiversity afforded by a wetland that is being considered as a site for an aquaculture facility. Some conflicts can be addressed through technology or improved management, but many are resolved only through economic incentives and market mechanisms, changes in how society values a resource, or development of environmental policies or regulations by governments or other institutions.

AQUACULTURE ENERGY

Aquaculture, as is true of agriculture in general, is a process whereby solar and fossil fuel energy are transferred from the environment into the culture system and converted into food energy. Although this concept oversimplifies the process because it ignores non-energy aspects of food quality, it is conceptually useful because most inputs to food production can be expressed in energy units, thereby making it possible to compare resource use among various aquaculture and agriculture production systems. For example, water use can be converted to energy units by quantifying the energy used to pump water, as well as the energy required to build and install pumps, pipes, and other infrastructure related to the water supply.

Quantifying energy use across various aquaculture production activities (production of juveniles, facility construction, feeds and feeding, labour, processing, and so on) also provides a means of identifying production inefficiencies where alternative practices may lead to energy and costs savings. The two primary energy sources for aquaculture are solar radiation and fossil fuels. No aquaculture system relies entirely on solar radiation, although sunlight is ultimately the major source of energy for all agriculture because plant material produced in photosynthesis is the base for nearly all food chains. But even in extensive cultures of seaweed and other aquatic plants where solar energy fully supports crop growth, fossil fuels are expended in functions such as harvesting and processing. Likewise, certain types of shellfish aqua-culture can be highly energy efficient because filter-feeders grow by consuming phyto-plankton swept past the beds or farms by tidal currents, but fossil fuel expenditures are made in the hatchery, harvesting, and processing phases of production and human labour is required during production, harvesting, and processing. Traditional pond aquaculture, as originally practiced in China, relied heavily on solar energy incident on the culture system to provide food, via photosynthesis, to support fish production.

Primary production was enhanced by fertilizing the system with plant nutrients derived from terrestrial plants, animal manures, or processing by-products of agricultural crops. Two or more fish species, usually carp, were cultured together (*polyculture*) to make efficient use of the variety of natural foods produced within the pond. As originally conceived, fish were locally consumed as fresh product, which reduced energy costs for transportation and eliminated energy costs for storage.

In the last half of the 20th century, global aquaculture transitioned from low-input systems relying heavily on solar radiation incident on the culture facility to systems requiring greater imports of energy from other sources. The goal of aquaculture shifted from low-intensity subsistence production of diverse species to intensive aquaculture focused on maximizing economic benefits through increased yields. As the goals of aquaculture changed, aquaculture systems increasingly relied on fossil fuels and appropriation of fixed solar energy from other ecosystems.

ENERGY INPUTS IN FISHERIES AND AQUACULTURE

Energy used in fisheries and aquaculture can be categorized as 1) ecosystem support; 2) direct energy inputs, and 3) indirect or embodied energy. *Ecosystem support* accounts for solar energy fixed in other ecosystems and then "imported" to the aquaculture system. An example of ecosystem support is the solar energy used in photosynthesis to produce soybeans and other plant feedstuffs that are used in manufactured feeds. Similarly, photosynthesis by marine phytoplankton is the base of the food web culminating in pelagic marine fish harvested for fishmeal and fish oil. Ecosystem support also includes energy used by natural processes to assimilate wastes produced during culture. For example, waste nitrogen and phosphorus produced during aquaculture may be discharged to a wetland where they are assimilated by plants. A certain amount of solar radiation and ecosystem area is needed to support the plant growth required to assimilate those nutrients. In general, aquaculture systems relying on manufactured feeds to support production (which includes much of finfish aquaculture in the United States) and those systems that discharge directly to public waters (net pens and flow-through systems) require a large ecosystem area outside the culture system to concentrate solar radiation into the chemical energy of feedstuffs and to process and assimilate wastes produced during culture.

Direct energy inputs are predominantly fossil fuels and labour used to provide most other resources needed in production. Examples include easily measured farm-level inputs such as energy used for pumping water and mechanical aeration. Other direct energy inputs are not so obvious. For example, a significant input to many types of aquaculture is fuel and labour used by farms and mills to produce and then process plant feedstuffs into aquafeeds. Another important set of energy inputs is the fuel, electrical energy, and labour used by fishing fleets and fish reduction plants to provide fishmeal. Direct energy inputs are highest for culture systems that rely on high-quality aquafeeds and in recirculating aquaculture systems where waste processing requires fossil fuel input.

Indirect energy inputs, also called *embodied energy inputs,* account for the energy needed to construct and maintain tanks, nets, ponds, buildings, pumps, aerators, boats, and other fixed assets needed to produce the aquaculture crop. Depending on the extent to which energy inputs are accounted, indirect energy costs may even include the energy and labour needed to obtain the raw materials—such as steel, wood, and plastic—used to fabricate the asset. Indirect energy inputs are highest for intensive aquaculture facilities that have significant physical infrastructure, but indirect energy inputs typically represent only a small fraction of overall energy costs of intensive aquaculture production because of the overwhelming energy inputs associated with production of manufactured feeds.

Life-cycle Energy Assessment

Life-cycle assessment (LCA) is a scientific discipline that attempts to account for all environmental impacts of providing specific goods and services to society. Life-cycle assessment derives its name from the concept that all products have a "life" starting and ending at predefined points that set boundaries for the assessment. Product life in agriculture may, for example, begin with acquisition of raw materials (fertilizers, feedstuffs, etc.) and include production, processing, transportation, and so on. Assessments can be made on the basis of any impact of interest. For example, total contribution to greenhouse gases can be estimated for all activities involved in production and retirement of a particular product. Other commonly used impacts in LCA include eutrophication, ozone creation, ecotoxicology, carcinogen production, land use, and water use. One of the most useful, and common, bases for LCA is energy input over product life. Energy as a basis for LCA is appealing because it is the simplest and most intuitive "common currency" for comparing impacts of all activities involved in production, use, and retirement among various products, processes, or activities.

Studies using LCA methodology consistently show that the energy used to produce manufactured feeds dominates the energetics of many modern aquaculture production systems. For example, more than 75 per cent of the total energy cost of producing Atlantic salmon in net pens is used in procuring or growing feed ingredients and manufacturing the feed. The remaining energy inputs, in order of importance, were fuel and electricity used to operate the facility, embodied energy costs (manufacture, maintenance, etc.) associated with physical infrastructure, and energy used to produce smolts. Feed production dominates energy budgets of all aquaculture systems relying on manufactured feeds, regardless of species, and overall energy use per unit protein production decreases in aquaculture systems less reliant on manufactured feeds.

Energy use in Aquaculture

The energy efficiencies of various aquaculture systems span perhaps the widest range of any agricultural sector. Traditional carp polyculture in ponds fertilized with agricultural by-products lies at one end of the spectrum as one of the most energy-efficient food production systems ever devised. When energy efficiency is expressed on the basis of industrial energy input per unit of edible protein energy produced, traditional carp polyculture rivals even vegetable crops for energy efficiency, with energy input/output ratios approaching 1. Most agricultural activities have energy input/output ratios much greater than 1, meaning that energy inputs during production far exceed food protein energy output. Relative to traditional pond culture of carp, aquaculture systems that rely heavily on manufactured feeds and other ecosystem support functions lie at the other end of the spectrum, with energy input/energy output values exceeding 50.

Nonetheless, most modern aquaculture systems are generally comparable to terrestrial animal production systems with respect to energy efficiency. For example, input/output energy ratios for pond-grown tilapia and channel catfish are similar to those for several common animal production activities, such as eggs, poultry (broiler), and swine production. Likewise, input/output energy ratios for marine shrimp aquaculture, which are among the highest of the major aquaculture systems, are in the same general range as that for shrimp trawling.

Comparing energy use in aquaculture and other forms of agriculture is difficult—and sometimes misleading—because few energy analyses of aquaculture have been conducted. Further, studies may not include adequate information on the boundaries of the analysis, which will bias comparisons among systems. Depending on the intent of the analysis, an almost endless number of input functions can be included in an LCA for an agricultural activity. Most studies of energy use include only the most easily quantified inputs, such as direct electrical and fuel inputs for the most obvious production functions (such as feed manufacture, water pumping, aeration, and so on).

Life-cycle assessment of energy use can include postharvest functions such as processing, freezing, refrigeration, storage, transportation, marketing, waste treatment, and even household activities such as refrigeration, freezing, and cooking. Energy use in these activities apparently has not been assessed for aquaculture but may be an important part of the overall energy costs of delivering aquaculture products to a consumer's plate. For example, energy used in on-farm production of the United States food supply accounts for only about 20 per cent of the energy used to deliver food to the consumer's plate. Postharvest processing and transportation each consume about 15 per cent and household preparation accounts for more than 30 per cent of the total energy consumed.

Ultimately it will be economically and socially imperative to improve the energy efficiency of all aspects of the food-supply chain. However, it is possible that greater overall gains in energy savings can be made by improving the efficiencies of processing, transport, retailing, and even household storage and preparation than can be made by improving energy efficiency in the production sector. This may have particular relevance to aquaculture, where important products are produced only in certain regions (marine shrimp in the tropics; salmon in the north-temperate) and are stored and shipped long distances for consumption.

Land

Aquaculture uses land in two ways. First, aquaculture facilities occupy a defined area or space on land or in water. But facility area accounts for only a portion of the total land or water area needed to produce an aquaculture crop. Additional ecosystem area is needed to provide support or service functions. The two most important of those functions are food production and waste treatment.

Facility Area

The area occupied by aquaculture facilities in the United States varies widely. There are approximately 125,000 ha of commercial aquaculture ponds in the United States, with 50,000 ha devoted to channel catfish culture. Contiguous blocks of catfish ponds in Mississippi may cover several hundred hectares on individual farms. By comparison, the total United States production of Atlantic salmon is derived from only a few hundred hectares of net cages on the Atlantic and Pacific coasts.

The land or water area needed to produce a unit of aquaculture crop is inversely proportional to production intensity. Production intensity is a vague term that attempts to describe the nature of resource use in aquaculture production. A simple, and common, measure of intensity is crop yield per unit of resource use, such as land. The land or water area needed per unit production of aquaculture crop varies over more than two orders of magnitude. At one extreme are highly intensive water recirculating aquaculture systems, which are capable of annually producing 1 to 2 million kg of fish per hectare of culture unit. Rainbow trout production in raceways is about 10 times less intensive and therefore requires about 10 times the surface area per unit of fish production. Fish and shrimp production in ponds requires several hundred times the land area compared with intensive recirculating systems.

Differences in the relative sizes of aquaculture facilities (or, more simply, the intensity of the system) are not simply a function of one system being inherently more efficient than another. Rather, the area required to produce a unit crop yield depends on the extent to which food production and waste treatment are subsidized by ecosystems or processes external to the culture system. For example, much of the ecosystem support for traditional pond aquaculture is inherent in the system. Traditional aquaculture ponds used in Chinese carp polyculture function not only to confine fish, but also to provide an internal area for food production and waste treatment. Land requirements for pond aquaculture are therefore relatively large. In net-pen culture, on the other hand, the culture unit functions only to confine the crop. Food is produced and wastes are treated in separate, external ecosystems. As such, the relative area needed for a net-pen facility is much smaller than for ponds.

Ecosystem Support: Food

Many, if not most, aquaculture systems in the world rely on plant photosynthesis either within the culture unit (ponds) or in nearby waters (open-water molluscan shellfish culture) to produce most of the food that supports aquaculture production. In ponds relying on *autochthonous* (within-pond) food production, aquaculture yield is limited by the rate of primary production, which in turn is ultimately limited by the amount of solar radiation impinging directly on the culture unit. Such systems are usually fertilized with plant nutrients to make fullest use of incident solar radiation and often contain two or more herbivorous and omnivorous species to make efficient use of the wide variety of natural foods produced in fertilized ponds. Production in molluscan shellfish aquaculture depends on solar radiation over a wide area to produce natural foods that are swept past and captured by the filter-feeding organisms.

Production of herbivorous or omnivorous fish can be quite impressive (3,000 to 10,000 kg/ha per year, or more) in fertilized ponds. The food web in these systems is relatively simple, and solar energy captured by plants is transferred efficiently to fish at these lower trophic levels. To increase aquaculture productivity past that achievable in systems dependent only on autochthonous primary production, food produced outside the culture system must be imported. In some systems, that food may be low-quality organic matter that might otherwise be considered a waste product, such as agricultural byproducts or livestock manures. In many aquaculture systems worldwide—and in nearly all commercial systems in the United States—the *allochthonous* (from outside the system) organic matter added to increase per-area production consists of high-quality manufactured feeds made from plant (soybean and corn, for example) and animal (usually fish) meals. Production of feedstuffs requires land external to the culture unit.

Soybean meal, wheat middlings, cornmeal, and cottonseed meal are common plant products used in aquafeeds. Boyd *et al.* (in press) calculated land areas needed to procure terrestrial feed ingredients to produce 1 tonne of various aquaculture species by using typical aquaculture production data, feed composition, and information from the United States Department of Agriculture for average plant seed and meal yields. Channel catfish is an example of a species grown on feeds with high levels of plant materials (>85 per cent of the diet) and low levels of fishmeal and fish oil (<5 per cent of the diet). Producing 1 tonne of channel catfish requires about 0.4 ha of cropland to produce plant feedstuffs. Atlantic salmon are fed diets with relatively low amounts of plant ingredients (20 to 40 per cent) and high amounts of fishmeal and fish oil (60 to 80 per cent of the diet). Atlantic salmon production requires a land area of about 0.1 ha/tonne for plant feed ingredients.

Marine ecosystems are required to produce small pelagic fish for reduction to meal and oil for inclusion in manufactured feeds. The appropriated marine ecosystem area can be calculated from the primary productivity of the fishing area and the trophic level of the pelagic fish captured for reduction. Values range from less than 10 to more than 100 ha of marine ecosystem area per tonne of fishmeal produced. Assuming a value of 50 ha of marine area per tonne of fishmeal, and typical feeds and production conditions used by Boyd *et al.*, approximately 2.2 ha of ocean area is needed to provide fishmeal to grow 1 tonne of channel catfish and about 18 ha of ocean is needed to grow 1 tonne of salmon.

These values are indicative rather than definitive because marine areas needed for fishmeal production vary greatly depending on the type of fishmeal used, the productivity of the fishery, and feed conversion when fed to fish. For example, the value of 18 ha/tonne calculated here for net-pen cultured Atlantic salmon is higher than the value of 14.2 ha/tonne reported by Tyedmers and much lower than the value of 100 ha/tonne reported by Folke. The qualitative implications of this exercise are 1) providing terrestrial ingredients for fish feeds appropriates relatively small areas of land whereas providing fish-derived feedstuffs requires relatively larger ecosystem support areas and 2) total appropriated ecosystem area for food production in intensive aquaculture is reduced when animals, such as channel catfish, from lower trophic levels are cultured.

Ecosystem support is often expressed on the basis of hectare of support area per hectare of culture area. When expressed in this manner, the dependence on ecosystem support varies over four orders of magnitude, depending upon culture system and feed composition. Intensive net-pen culture of carnivorous fish, such as salmon, requires an ecosystem support area for food production that is more than 10,000 times the area of the culture system. Growing the omnivorous species, channel catfish, in ponds appropriates an area for procurement of feedstuffs that is approximately 10 times the area of the pond (recalculated from Boyd *et al.*, in press, using additional data on ecosystem areas needed for fishmeal). At the extreme, carp, tilapia, and other fish from lower trophic levels can be grown in ponds fertilized with agricultural wastes or byproducts, and do not require external ecosystem area for food production. Those systems have a value of 1.0 for the ratio of ecosystem support to culture area.

Open-water molluscan shellfish culture is unique in that animal growth depends on natural foods produced in marine or brackishwater ecosystems that are much larger than the culture area. Those foods are swept over the culture area by tidal currents and removed by the filter-feeding animals. In one context, mollusk culture "appropriates" food produced in an area much larger than the culture area, but this food is not cultivated and requires no additional input of resources past solar radiation and nutrients already present in the water. Natural food organisms consumed by mollusks and the nutrients that supported their growth are usually considered underutilized, and their removal from the ecosystem is seen as beneficial. Accordingly, no additional ecosystem support area should be assigned to the final grow-out phase of open-water mollusk culture.

Ecosystem Support: Waste Treatment

With the exception of molluscan shellfish culture, all aquaculture systems use resources procured from one ecosystem, concentrated, and then added to the aquaculture system, inevitably producing waste. In addition to land area for facilities and to produce food, ecosystem area is therefore required to assimilate those wastes. In recirculating aquaculture systems and ponds operated with long hydraulic residence times, significant quantities of waste produced during culture are treated within the facility and there is relatively little external area needed for waste treatment. On the other hand, much of the waste produced in raceway and net-pen culture is discharged directly to the outside environment.

The ability of the external ecosystem to assimilate those wastes may limit aquaculture production either by polluting the surrounding water to the point where animal welfare inside the facility is endangered (*self-pollution*) or by imposing regulatory constraints on the amount of waste that can be discharged. In addition to effects on aquaculture production, waste discharge into public waters may create societal externalities such as degraded water quality that limit options for use, water treatment costs, and other downstream impacts.

The ecosystem area needed for waste assimilation, expressed as hectares of waste treatment area per hectare of production facility, varies over at least two orders of magnitude, depending on the type of production system. At one extreme are aquaculture ponds with low to moderate stocking and feeding rates that can, in theory, be operated for many years without intentional water exchange to remove wastes.

Natural biological, chemical, and physical processes active inside the pond remove or transform wastes at rates adequate to prevent year-to-year accumulation of potential pollutants. In theory, no outside ecosystem support area is needed to treat wastes produced during culture and, therefore, the ratio of land area for waste treatment divided by facility area is one. In other words, the pond functions as its own waste treatment facility. In practice, of course, ponds must be occasionally drained and some overflow is inevitable during periods of heavy precipitation. Nevertheless, for ponds operated with long hydraulic residence times, more than 90 per cent of the waste organic matter, nitrogen, and phosphorus produced during culture is assimilated inside the pond before water is discharged.

Internalizing waste treatment imposes a relatively high direct land cost to pond aqua-culture, and this cost is evident in the large size of pond facilities. Extensive land use is the result of the pond functioning as both an animal confinement area as well as a waste treatment facility. For example, more than 95 per cent of the total area of a catfish pond functionally acts as a photosynthetic waste-treatment lagoon and less than 5 per cent of the total area serves as a "fish-holding" area. In effect, more than 95 per cent of the land and construction costs for ponds can be assigned to waste treatment functions, and the relatively large land area occupied by a pond aquaculture facility is a price the farmer pays for treating wastes on site rather than discharging the waste to public waters. Engle and Valdarrama explored other costs (such as aeration, labour, etc.) associated with internalizing waste treatment in channel catfish ponds and calculated that almost 30 per cent of the total cost of producing channel catfish can be ascribed to internal waste treatment processes. Land area requirements and other costs resulting from internalizing waste treatment in ponds limit profitable culture only to areas where large tracts of flat land are available at a reasonable price.

Aquaculture production in ponds is therefore limited by the finite capacity of the pond ecosystem to treat wastes produced during culture. Further intensification of production is possible only if wastes are treated external to the culture unit, usually by discharging wastes to public waters. Accordingly, ecosystem area outside the culture facility is used to assimilate wastes, and in most instances the cost of that treatment is borne by society, not the aquaculturist.

Relatively few studies have been conducted to determine the external ecosystem areas needed to treat aquaculture wastes produced by aquaculture in raceways or net pens. Furthermore, requirements will vary depending on the hydrology and biology of the ecosystem into which wastes are discharged. Based on the few studies conducted, it appears that ecosystem support areas between 100 and 300 times the facility area are needed to treat wastes produced from cage and net-pen fish culture. These values for ecosystem waste assimilation area are somewhat larger, but still within the same approximate order of magnitude as the ratio of waste treatment area to fish-holding area in the catfish ponds. This is not a coincidence, because the same biological and physicochemical processes are responsible for waste treatment whether the water is inside a pond or in the lake, stream, or estuary in which cages are suspended. Areal requirements for waste treatment should therefore be of similar magnitude. The difference, of course, is that the ecosystem area needed for waste treatment in ponds is, for the most part, inherent in the production system, whereas waste treatment for cage and net-pen culture is external to the system.

Mollusks generate wastes in the form of feces and pseudofeces that are deposited in sediments around shellfish culture areas. Those wastes may impact the local environment and ecosystem area is required to assimilate those wastes.

However, on a larger scale, mollusks are net consumers of particulates and nutrients, so in a sense, the local ecosystem area needed to assimilate wastes produced in the shellfish-growing area is less than the overall ecosystem area that would have been required to assimilate the waste in the absence of the filter-feeding shellfish. The concept of ecosystem support required to treat wastes is difficult to apply to open-water molluscan shellfish aquaculture.

ENERGY REQUIREMENTS

The energy needs for maintenance and activity must be satisfied before any growth can occur. Feeding levels must be high enough to supply maintenance needs and still have energy remaining for growth. Digestion efficiency in fish decreases as feeding level is increased. The problem becomes one of finding the feeding level at which the increased efficiency of energy utilization at a high feeding rate is balanced by the lower efficiency of digestion at the higher feeding rate.

ENERGY DISTRIBUTION IN RELATION TO FEEDING LEVEL

The distribution of dietary energy intake in relation to feeding level in fish. Basal or standard metabolism in fish is relatively constant under constant environmental conditions. It can change with changes in temperature and fish size among other factors. The energy expended on voluntary activity usually increases somewhat with increasing feeding level. Starved fish are less active than well-fed fish but there is always some expenditure of energy for activity. The heat of nutrient metabolism is proportional to the level of feeding. The energy excreted in urine and gill excretions is also a function of feeding level. The amount remaining for growth is zero at maintenance feeding and becomes proportionately greater as feeding level is increased, until it is balanced by the decreased efficiency, of digestion.

MAINTENANCE ENERGY

All of the energy lost due to standard metabolism, heat of nutrient metabolism and physical activity appears as heat. The maintenance requirement can be determined by measuring the heat produced. The heat production can be measured directly in a calorimetre or it can be estimated by measuring oxygen and applying the appropriate heat equivalent. The factor most commonly used is 3.42 kcal/mg O_2. This factor is largely an extrapolation of data for mammals and has not been directly measured in fish. The heat equivalent of oxygen also varies with the type of substrate being oxidized. Maintenance energy can also be estimated by measuring energy loss during starvation.

ENERGY COST OF GROWTH

It has been shown in mammals that the cost of growth is fairly constant after maintenance energy is subtracted from ME fed. This probably holds true for fish, but it has not been experimentally determined. More research is needed in this area.

FACTORS THAT ALTER ENERGY NEEDS

There are several factors which can alter the energy requirements of fish. Feeding rates should be adjusted to compensate for these factors to avoid overfeeding, but still providing sufficient energy for optimum growth.

- *Temperature*: As environmental temperature declines homeotherms must increase their metabolic rate to compensate for the additional heat loss if they are to maintain a constant body temperature. Most freshwater fish do not attempt to maintain a body temperature which is different from the environment. As water temperature declines, body temperature of the fish declines and metabolic rate is reduced. The low metabolic rate at low temperatures enables fish to survive for long periods under ice where little food is available.

There is considerable species difference in metabolic adaptation to environmental temperature changes. Each species seems to have a preferred temperature at which it functions most efficiently. If temperature gradients exist, the fish will seek the most favourable temperature. Usually this is the temperature at which the difference between maintenance requirement and voluntary food intake is greatest and at which optimum efficiency of growth occurs.

- *Water Flow*: Energy which is used for physical activity is not available for growth. Fish which are forced to swim against a strong current are expending energy which would otherwise be used for growth. However, still water allows stratification and the accumulation of waste products. Fish rearing facilities should be designed to obtain maximum use of water without undue stress on the fish.
- *Body Size*: Small animals produce more heat per unit weight than do large animals. Small fish should be fed a higher percentage of body weight than large fish. In mammals the metabolic rate is proportional to the three-fourths power of body weight. The exponent applicable to fish has been reported from 0.34 to 1.0. The factor $W^{0.8}$ usually used. Obviously more work is needed in this area. Work at the Tunison Laboratory of Fish Nutrition has indicated that rainbow trout from 1.0 to 4.0 g in weight have a metabolic rate proportional to $W^{1.0}$. Fish from 4.0 to 50.0 g in weight have metabolic rates proportional to $W^{0.63}$.

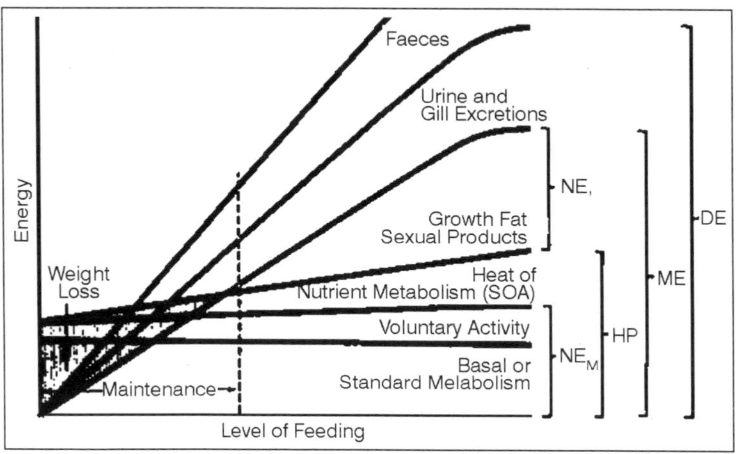

Fig. Distribution of Dietary Energy Intake in a Growing Fish at Various Levels of Feeding

- *Level of Feeding*: The level of feeding also has an effect on the energy expenditure of fish. This becomes important in design of fish rearing facilities. Dissolved oxygen is usually the first limiting factor in fish rearing. The oxygen consumption increases recently after feeding due to the physical activity of feeding and the heat of nutrient metabolism. Facilities must be designed with adequate safety margins. The oxygen required per unit weight of feed also varies with feeding level, being higher at maintenance level when all the food is oxidized than at higher feeding levels when much of the energy is stored as growth.
- *Other Factors*: Several other factors can contribute to high energy requirements. Anything which makes the fish uncomfortable increases physical activity and reduces growth. Crowding, low oxygen and waste accumulation are some of these factors.

BODY LIPID COMPOSITION AND DIETARY LIPID REQUIREMENTS

Detailed information is still lacking on the dietary lipid requirements of many species of fish, but there is an abundance of information on the fatty acid composition of fish oils. Information on the lipid composition of fish can be used to make some guesses about dietary lipid requirements. Linolenic acid resulted in some sparing action and growth promotion in rats, and fatty acids of the w 6 EFA prevented all of the EFA-deficiency symptoms.

Research with homeothermic land-dwelling animals showed that the w 6 series of fatty acids are the "essential fatty acids", while the w 3 series are considered to be non-essential or only have a partial sparing action on EFA-deficiency. The w 6 series of fatty acids have been shown to be essential to enough species that it began to become accepted that these are the essential fatty acids for all animals. It was assumed by many that fish also required w 6 fatty acids. Many researchers began by supplementing fish diets with vegetable oils, such as corn, peanut, or sunflower oil, which were rich in linoleic acid. The main sympton observed during the development of EFA deficiency in chinook salmon fed fat-free diets was a marked depigmentation that can be prevented by addition of 1 per cent trilinolein, but not by 0.1 per cent linolenic acid. Although the w 6 fatty acids are considered to be essential, one of the general characteristics of fish oils is the low levels of w 6 series fatty acids and the higher levels of w 3 type fatty acids. There is evidence that polyunsaturated fatty acids of the w 3 series, which are present in relatively large concentrations in fish oil, play the role of essential fatty acid for fish.

Polyunsaturated Fatty Acid

When a test diet containing 13 per cent corn oil and 2 per cent cod-liver oil was fed to rainbow trout, subsequent deletion of the cod-liver oil from the diet depressed growth and produced some kidney degeneration which might be attributed to a lack of sufficient w 3 PUFA present in significant quantities in cod-liver oil.

Dietary fish oil is superior to corn oil in promoting growth of rainbow trout and the yellow-tail. Dietary linolenic acid or ethyl linolenate gives a positive growth response for rainbow trout which may be attributed to a dietary requirement for w 3 fatty acids.

Essential Fatty Acid Requirements

One of the most widely accepted theories explaining the presence of such high levels of 20:5w 3 and 22:6w 3 fatty acids in fish oils is related to the effect of unsaturation on the melting point of a lipid. The greater degree of unsaturation of fatty acids in the fish phospholipids allows for flexibility of cell membrane at lower temperatures.

The w 3 structure allows a greater degree of unsaturation than the w 6 or w 9. This theory is consistent with the fact that cold water fish have a greater nutritional requirement for w 3 fatty acids, while the EFA requirement of some warm water fish can be satisfied by a mixture of w 6 plus w 3.

Rainbow Trout

Rainbow trout, a cold water fish, requires w 3 fatty acids as EFA in the diet. The EFA requirement can be met by 1 per cent 18:3w 3 in the diet. Inclusion of 18:2w 6 in the diet may result in some improvement in growth and feed conversion compared to EFA deficient diets; however, the w 6 fatty acids will not prevent some EFA deficiency symptoms such as the "shock syndrome". Although it is clear that rainbow trout require w 3 fatty acids, it remains to be shown conclusively whether some dietary level of w 6 fatty acid is essential.

Dietary 18:2w 6 or 18:3w 3 were readily converted to C-20 and C-22 PUFA of the same series, and 18:3w 3 or 22:6w 3 had similar EFA value for rainbow trout. Either 20:5w 3 or 22:6 w 3 is superior to 18:3w 3 in an EFA value for rainbow trout, and the former two fatty acids in combination are superior to either alone.

This is consistent with data for mammals, where 20:4w 6 has higher EFA value than 18:2w 6. The superior nutritional value of C-20 and C-22 carbon w 3-PUFA is further supported by the excellent growth promoting effects of dietary fish oils such as pollock liver oil and salmon oil for rainbow trout.

Channel Catfish

One of the most important warm water fish in North America is the channel catfish. The quantitative EFA requirement of the catfish has not yet been determined. However, the evidence is convincing that the ω 3 requirement is not as great as that of rainbow trout. Analysis of fatty acids of lipids from catfish purchased at five processing plants showed very low levels of 20:4ω 6, 20:5ω 3, and 22:6ω 3; 0.8 – 5.5, 0.2 - 1.3, and 0.6 - 6.1 per cent of the total fatty acids, respectively.

It was shown that corn oil added to a semipurified casein based diet initially resulted in a positive growth response and protein sparing, but later growth inhibition was observed. The apparent repressive effects of corn oil may be due to its 18:2 ω 6 content since 20:5 ω 3 and 22:6 ω 3 present in menhaden oil had no apparent detrimental effects. The growth suppressing effects of 18:2 ω 6 were also noted when 3 per cent corn oil was added to 3 per cent beef tallow and 3 per cent menhaden oil. The growth suppression caused by unsaturated fatty acids does not appear to be limited to ω 6 fatty acids. Linseed oil in the diet of catfish resulted in growth suppression similar to that caused by corn oil compared with dietary beef tallow, olive oil and menhaden oil.

The Common Carp

The common carp is much clearer than that for the channel catfish. This fish has an EFA requirement for both 3 and ω 6 fatty acids. The best weight gains and feed conversions are obtained in fish receiving a diet containing both 1 per cent 18:2 ω 6 and 1 per cent 18:3 ω 3.

With the carp, 20:5 ω 3 and 22:6 ω 3 at 0.5 per cent of the diet are superior to 1 per cent of 18:3u3. Carp fed a fat-free, or EFA deficient, diet incorporated high levels of 20:3 ω 9 in their lipids, especially in the phospholipids.

The Eel

The eel another warm water fish, has a requirement for both ω 3 and ω 6 fatty acids. Corn oil and cod liver oil in a 2:1 mixture are most favourable for the growth of eels. The eel requires ω 6 and ω 3 in the same proportion as the carp, but at a lower level in the diet; namely, 0.5 per cent of each, rather than 1.0 per cent of each PUFA.

The Plaice

The plaice becomes depleted of both ω 3 and ω 6 PUFA when fed a fat-free diet. The addition of 12:0 and 14:0 to the diet result in the synthesis of saturated and monoenoic fatty acids of chain lengths up to C18; however, increased levels of 20:3 ω 9 noted in trout and mammals have not been reported in plaice. Plaice fed dietary 18:2 ω 6 and 18:3 ω 3 will not produce significant amounts of 20:4 ω 6, 20:5 ω 3, or 22:6 ω 3.

The Turbot

The growth of turbot is much better with ω 3 PUFA than with ω 6 or saturated fat in the diet. The turbot also appears to be unable to convert dietary 18:2 ω 6 to 20:4ω 6 when fed corn oil, or to convert endogenous 18:1ω 9 to

20:3ω 9 when fed the EFA deficient diet. Although it appears to have an EFA requirement for w 3 fatty acids such as are present in cod liver oil, this requirement is not satisfied by 18:3ω 3.

The chain elongation and desaturation of 18:lω 9, 18:2 ω 6, or 18:3ω 3 has been found to be very limited in turbot compared to the rainbow trout where 70 per cent of the 18:3w 3 was converted to 22:6ω 3. The required level of long-chain ω 3 fatty acids for turbot is at least 0.8 per cent of the diet.

The Red Sea Bream

The red sea bream grows better when the dietary lipid is of marine origin rather than a vegetable oil. The EFA requirement of the red sea bream is not satisfied by either linoleic acid of corn oil or supplemented linolenate.

A mixture of 20:5ω 3 and 22:6ω 3 supplemented to the corn oil diet has been shown to be effective in improving growth and condition of these fish. Thus, even in warm water, marine fish seem to require not just ω 3 fatty acids but 0)3 fatty acids of 20 to 22 carbon-chain length. A direct correlation between feed efficiency and the 18:1 level in the lipids of the red sea bream has been postulated.

Other Species

Among warm water marine fish, mullet and fundulus possess the ability to chain, elongate, and desaturate 18:2ω 6 or 18:3ω 3 PUFAs. The process is, however, inhibited in fundulus by high levels of these PUFAs of 18:2ω 6 or 18:3 ω 3 in the diet.

FATTY ACID METABOLISM

It appears that high levels of 18-carbon w 6 or ω 3 fatty acids inhibit the synthesis and metabolism of 18:lω 9. It is interesting to note that the channel catfish, which also exhibits negative growth response to dietary 18:2ω 6 or 18:3ω 3, incorporates very high levels of 18:1 into its body lipids. The inclusion of either 18:2ω 6 or 18:3ω 3 in the diet reduces the levels of 18:1 fatty acids in body lipids. A similar reduction has also been observed in red sea bream liver phospholipid when either of the PUFAs is added to the diet.

The competitive inhibition of chain elongation and desaturation of members from one series of fatty acids for members of another series is well established, with ω 3 > ω 6 > ω 9 being the usual order of potency for inhibition.

The pathways of fatty acid metabolism have been reviewed by Mead and Kayama. Fish are able to synthesize, de novo from acetate, the even-chain, saturated fatty acids. Radio tracer studies have shown that fish can convert 16:0 to the ω 7 monoene and 18:0 to the ω 9 monoene. The ω 5, ω 11 and ω 3 monoenes are proposed based on the identification of these isomers in the monoenes of herring oil. Fish are unable to synthesize any fatty acids of the ω

6 and u3 series unless a precursor with this w structure is present in the diet. Fish are able to desaturate and elongate fatty acids of the ω 9, ω 6, or ω 3 series. There is competitive inhibition of the elongation desaturation of fatty acids of one series by members of the other series.

The ω 3 fatty acids are the most potent inhibitors, the w 9 are the least. The ability to elongate and desaturate fatty acids is not the same in all species of fish, as was noted earlier. The turbot was able to desaturate and elongate only 3-15 per cent of 18:1ω 9, 18:2ω 6, or 18:3 ω 3, when given the C^{14} labelled fatty acid; in the rainbow trout, 70 per cent of the label from 18:3ω 3 was found in 22:6 ω 3. The essential fatty acids are not unique in their ability to supply energy. The β -oxidation of fatty acids in fish is basically the same as in mammals. The EFA and saturated and monoenoic fatty acids are all equally utilized by fish for energy production.

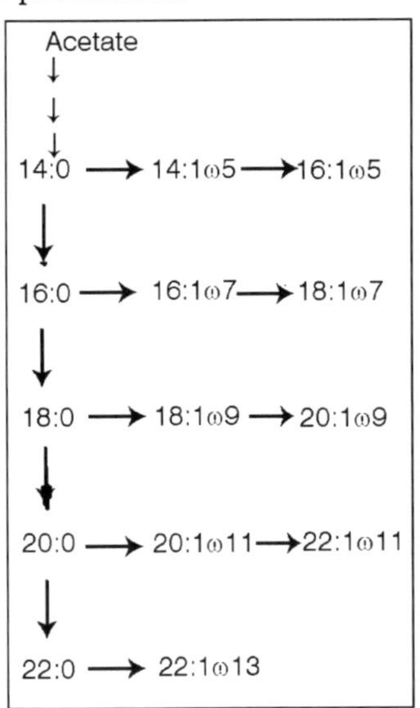

Fig. Polyunsaturated Fatty Acids

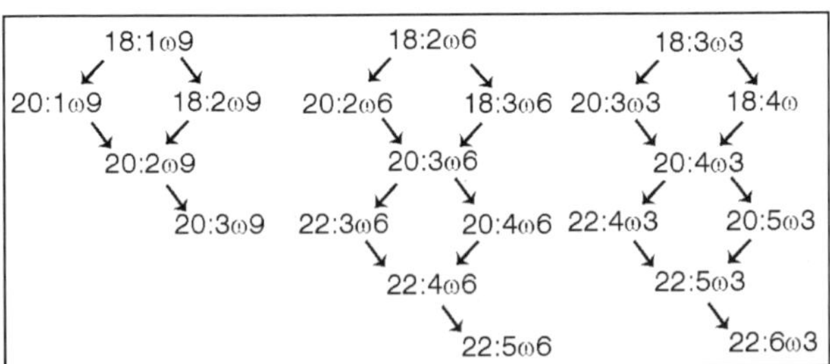

Increased swelling rates of liver mitochondria occur in rainbow trout fed diets deficient in ω 3 fatty acids. It is possible that EFA plays an important role in the permeability as well as the plasticity of membranes. The role of ω 3 fatty acids in membrane permeability may be one of the factors accounting for differences in content of this family of fatty acids between freshwater and marine fish. Fish mitochondria with high levels of the ω 3 PUFA and very low levels of ω 6 fatty acids are very similar to mammalian mitochondria with respect to cytochrome content, β -oxidation of fatty acids, operation of the tricarboxylic acid cycle, electron transport, and oxidative phosphorylation. The ω 3 PUFA may play the same role in fish that the ω 6 fatty acids play in rats. The EFA play another role in the mitochondria. In addition to their importance in membrane structure, the EFA are important in some enzyme systems.

Unsaturated fatty acids play an important role in the transportation of other lipids. It has been repeatedly shown that feeding PUFA will lower the cholesterol levels in animals with above-normal blood lipid and cholesterol levels. Fish oils are more effective in lowering cholesterol levels than are most dietary lipids.

The major portion of the fatty acids absorbed across the intestinal mucosa are transported as protein-lipid complexes stabilized by phospholipids. The low body temperature in fish probably results in a greater importance for unsaturation in transport of lipids than in homeothermic animals.

NEGATIVE ASPECTS OF LIPIDS IN NUTRITION

The requirement by fish for PUFA of the ω 3 series creates problems with respect to feed storage. These types of fatty acids are very labile on oxidation. The products of lipid oxidation may react with other nutrients such as proteins, vitamins, etc., and reducing the available dietary levels or the oxidation products may be toxic. The effect of oxidized lipids on dietary proteins, enzymes and amino acids have been demonstrated.

The use of oxidized menhaden oil in the diets of swine and rats caused decreased appetite, reduced growth, yellowish-brown pigmentation of depot fat, and decreased haemoglobin and haematocrit levels. The negative effects of the oxidized fish oils were reversed by the addition of alpha-tocopherol acetate or ethoxyoquin to the diet.

Much of the use of vegetable oils in fish diets in the 1950s and 1960s might, in part, have been based on their greater stability in prepared diets. It has been demonstrated that rancid herring and hake meals in fish feeds caused dark colouration, anaemia, lethargy, brown-yellow pigmented liver, abnormal kidneys, and small gill clubbing in chinook salmon. The symptons can be alleviated by addition of alpha-tocopherol to the diets containing rancid fish meals. The addition of vitamin E would prevent the toxic or negative effects of adding 5 per cent highly oxidized salmon oil to the diet of rainbow trout. This same sparing effect of alpha-tocopherol can also apply to rancid carp feed.

The positive nutritional value of w 3 fatty acids in fish lipids for fish feeds can become a negative factor if adequate care is not taken in the preparation and storage of feeds. Only fresh oils with low peroxide values should be used in feeds. Fish feed ingredients such as fish meals should be protected against oxidation. The level of vitamin E added to the diet should be increased as the PUFA level is increased. The finished feed, if possible, should be stored in air tight containers at reduced temperatures with minimum exposure to UV radiation and other factors accelerating the rate of lipid oxidation. The problems of rancidity or antioxidation of lipids in fish feeds should not be ignored.

REQUIREMENT FOR WATER BY AQUACULTURE

The requirement for water by aquaculture varies depending upon the type of culture system. Flow-through systems (such as raceways) require large volumes of water per unit fish produced. For example, water volume use in raceway culture of rainbow trout in the United States is approximately 98 m^3/kg. The most common water supplies for flow-through systems are artesian springs or surface waters diverted from streams or rivers. Pumping water from wells or other sources is too expensive for large-scale commercial aquaculture and is seldom used except in small public hatcheries or in commercial hatcheries that are used for only part of the year (as in channel catfish farming, for example; Tucker 2005).

Water flowing into culture units provides dissolved oxygen and water flowing out carries away metabolic wastes, such as carbon dioxide, ammonia, and fecal solids. Flow-through systems are usually configured so a given parcel of water is used more than once as it flows through a series of culture units. However, without treatment of the water to add oxygen or remove products of fish metabolism, there is a limit to the number of raceway segments that can reuse the same parcel of water.

Although flow-through systems require large volumes of water, they do not consume water unless it is pumped from aquifers. Except for negligible amounts of water lost to evaporation and fish harvest, inflow equals outflow. Boyd estimated that consumptive water use in a typical trout flow-through facility is about 0.03 m^3/kg of production, or less than 0.05 per cent of total water use. The common interpretation of water conservation is, therefore, meaningless in flow-through aquaculture because essentially no water is lost.

However, efficiency of water use can be improved by using technologies that increase fish production per unit water flow. Although modern waste-management technologies used in flow-through aquaculture can remove a large proportion of the wastes added to water during culture, it is inevitable that some waste is discharged in flow-through system efflu-ents. In that regard, flow-through systems "use" water because downstream ecosystems are impacted and "appropriated" for waste treatment.

Water use in net-pen culture is similar to that in raceways. Adequate water quality within net pens depends on the exchange of water from outside the unit by tidal currents or river flow.

In essence, net pens are simply flow-through systems constructed in water instead of on land.

Like flow-through systems, large quantities of water must flow through the culture units to provide adequate dissolved oxygen and remove wastes, but net-pen facilities do not consume water. Most of the waste produced during culture is discharged directly into public, multiple-use water bodies and, as with raceways, net pens appropriate ecosystem area for waste treatment. The impact of appropriating surrounding areas for waste treatment can be minimized through proper site selection.

In contrast to flow-through and net-pen systems, many recirculating aquaculture systems are operated with very low water exchange rates. Water within culture units is reused many times by cycling water through mechanical and biological treatment processes to add dissolved oxygen and remove solids, ammonia, and dissolved carbon dioxide. In the extreme, water is used only to replace evaporative losses (which are small) and to replace water lost when concentrated waste solids are discharged from the system. Total and consumptive water use in recirculating can be less than 0.1 m^3/kg of production.

Water use in ponds is functionally similar to that in recirculating systems in that water quality is maintained by mechanical and biological processes. However, in recirculating systems, treatment takes place in discrete units, such as screens, filters, or settling basins, whereas in ponds the processes are inherent parts of the ecosystem. Water use in ponds varies over a much wider range that for any other aquaculture system, depending primarily on the frequency of intentional water exchange (flushing) and pond drawdowns for harvest. At one extreme, water exchange in some ponds is so frequent that the systems operate hydrologically more like flow-through systems than ponds. In the 1980s and 1990s, marine shrimp ponds were routinely operated with daily water exchange rates of 10 to 20 per cent of pond volume, and total water use (40 to 80 m^3/kg of shrimp produced) approached that of flow-through systems. Water used for water exchange is pumped from bays or estuaries and discharged back into the same water body, so only water lost to evaporation and seepage is used consumptively. More important than water use alone, however, water exchange generates large volumes of waste, increases the risk of pathogen transfer to and from the outside environment, and increases production costs associated with pumping.

Eliminating or reducing water exchange by using water-reuse technologies can reduce total water use in shrimp ponds to less than 2 m^3/kg. Total water use in channel catfish farming (as an example of pond aquaculture in general) varies greatly depending on water source and phase of culture.

Ponds may be filled with either watershed run-off or from wells, and consumptive use is much greater for ponds using groundwater. Consumptive use also varies among the different phases of catfish culture because fingerling ponds are drained and refilled each year whereas brood ponds and foodfish ponds are operated for several years without draining.

As such, more water is used for the fingerling phase because water drained each year must be replaced. Average consumptive water use in the foodfish-production phase is relatively low (less than 2 m^3/kg) because ponds are operated for long periods without draining.

However, actual water use to produce foodfish must include water used in fingerling and brood phases of culture. When annual water use in the three culture phases are added and then divided by total annual foodfish production, average consumptive water use for pond-grown catfish in the southeastern United States is approximately 3 m^3/kg. Consumptive water use is much higher in ponds than for other culture systems, and significant opportunity exits for water conservation. Considerable attention is therefore given to hydrology and water conservation.

The relatively high economic value of water used in aquaculture (also called the *consumptive water value index*) can be shown by comparing consumptive water use and gross economic value for aquaculture and irrigated terrestrial crops grown in the same region. Using data from the southeastern United States, channel catfish aquaculture requires more water than irrigated cotton, corn, and soybeans, but an amount comparable to rice. However, the value of various crops per unit volume of water used is much greater for catfish than other crops. If this is interpreted as an index of water use economic efficiency, catfish aquaculture is a more efficient user of water than irrigated row crops.

FEED AND GOAL OF AQUACULTURE

The goal of aquaculture is to manage a body of water so that it will produce more fish, crustaceans, mollusks, or other animals than it would without management. Except for molluscan shellfish culture, this is accomplished by concentrating resources to provide food for the animal. There are two ways to provide food for aquatic animals:

- Fertilize an outdoor system to enhance natural food production and
- Grow foods outside the culture unit.

The energetics and land-use issues associated with these two approaches.

In the context of the ecosystem-level support for food production, the common aqua-culture systems used in the United States range from open-water mollusk culture, with no intentional resource input to enhance food supply, to net-pen and flow-through system culture of salmonids, where animal growth is totally dependent on resource expenditures outside the culture unit. Between those extremes is a continuum of food-supply strategies.

In addition to molluscan shellfish culture, several other species are grown with relatively low resource input (aside from labour). For example, crawfish ponds operated at low intensity depend on development of a detrital food web based on natural plant forage.

Some baitfish and ornamental fish ponds are also operated at relatively low intensity and are fertilized to promote an abundant supply of natural foods. However, more than 95 per cent of all commercial finfish aquaculture and 70 per cent of total aquaculture production in the United States depends on manufactured feeds. Worldwide, manufactured feeds support approximately 40 per cent of all animal aquaculture production.

Manufactured feeds are formulated to provide all essential nutrients and energy needed to promote rapid growth of healthy animals. Diet composition varies widely depending on the nutritional requirements of the species cultured, cost of feedstuffs, and the relationship of feed formulation to body composition, disease resistance, and waste production.

Feeds are typically formulated from a mixture of animal and plant feedstuffs. The major animal feedstuff in most aquafeeds is fishmeal derived from pelagic marine feed fish such as anchovy, menhaden, and herring.

Other animal feedstuffs include poultry and beef processing by-products. The primary plant feedstuffs are meals or by-products derived from soybeans, wheat, corn, canola and rapeseed, rice, sorghum, cotton seed, or peanuts. Diets may also contain various animal and plant fats or oils to increase feed energy content, provide essential fatty acids, and reduce dust or fines produced during feed handling. Fish oil is presently an indispensable ingredient in diets for salmonids and other carnivorous finfish and is a coproduct of fishmeal production. Feeds may also contain supplemental vitamins and minerals, binding agents, or pigments. Summaries of feed formulation for most aquaculture species are provided in Lovell and Webster and Lim.

LIPIDS AND FATTY ACIDS

Lipids are the generic names assigned to a group of fat soluble compounds found in the tissues of plants and animals,: and are broadly classified as:

- Fats,
- Phospholipids,
- Sphingomyelins,
- Waxes,
- Sterols.

Fats are the fatty acid esters of glycerol and are the primary energy depots of animals. These are used for long-term energy requirements during periods of extensive exercise or during periods of inadequate food and energy intake. Fish have the unique capability of metabolizing these compounds readily and, as a result, can exist for long periods of time under conditions of food deprivation. A typical example is the many weeks of migration by salmon in their return upstream to spawn; stored lipid deposits are burned for fuel to enable body processes to continue during the strenuous journey. Phospholipids are the esters of fatty acids and phosphatidic acid. These are the main constituent lipids of cellular membranes allowing the membrane surfaces to be hydrophobic or hydrophylic depending on the orientation of the lipid compounds into the intra or extracellular spaces.

Sphingomyelins are the fatty acid esters of sphingosine and are present in brain and nerve tissue compounds.

Waxes are fatty acid esters of long-chain alcohols. These compounds can be metabolized for energy and to impart physical and chemical characteristics through the stored lipids of some plant and several animal compounds.

Sterols are polycyclic, long-chain alcohols and function as components of several hormone systems, especially in sexual maturation and sex-related physiological functions.

Fatty acids can exist as straight chain or branch chain components; many of the fish fats contain numerous unsaturated double bonds in the fatty acid structures. A short bond designation for. fatty acids will be used throughout where the w number identifies the position of the first double bond counting from the methyl end. Linolenic acid would be written 18:3w 3. The first number identifies the number of carbons; the second number, the number of double bonds; and the last number, the position of the double bonds.

Many reviews of fish nutrition have been published which contain information on lipid requirements. Most work on lipid requirements of fish has been with salmonids. Rainbow trout have an essential fatty acid (EFA) requirement for the linolenic of w 3[1] series rather than for linolenic or w 6 as required by most mammals. The main emphasis on lipid requirements has been on EFA and on the energy value of lipids.

FATTY ACID COMPOSITION

Environmental Influences

Salinity

The difference between fatty acid compositions of marine and freshwater fish has been noted by several authors. Although these fish lipids are higher in w 3 fatty acids, it is clear that freshwater fish have higher levels of w 6 fatty acids than marine species. The average w 6/w 3 ratios are 0.37 and 0.16 for freshwater and marine fish, respectively. Fish in general contain more w 3 than w 6 polyunsaturated fatty acids and should have a higher dietary requirement for w 3 PUFA; thus the dietary EFA requirement of marine fish for w 3 PUFA may be higher than that of freshwater fish. The same type of difference in the w 6/w 3 ratio between freshwater and seawater is seen when some species of fish migrate from oceans to streams or vice versa. The PUFA ratio of sweet smelt changes drastically in only one month as they migrate from the sea to a freshwater river. A similar but reverse change occurs in the masu salmon as they migrate from freshwater to seawater. Even within the same species of fish, the salinity of the water seems to cause a dramatic change in the fatty acid pattern.

The difference between marine and freshwater fish may be due simply to differences in the fatty acid content in the diet or it may be related to a specific requirement in fish related to physiological adaptations to the environments. The phospholipids are generally considered to be structural or functional lipid, being incorporated to a large extent in the membrane structure of cell and subcellular particles.

The triglycerides are more often storage lipids and reflect the fatty acid composition of the diet to a greater extent than do the phospholipids. The fatty acid compositions of the triglyceride and phospholipid fractions of fish lipids are presented. It can be seen that the effect of changing environment on the fatty acid composition of the phospholipid is as great in the case of salmon, and considerably greater in the case of the sweet smelt, than it is on the triglyceride composition. Rainbow trout on diets containing either corn oil, which is high in w 6 but low in w 6 PUFA, showed a higher mortality and growth reduction in seawater than in freshwater over the twelve-week feeding period.

Temperature

There are several other factors besides the salinity of the water which affect the fatty acid composition and especially the PUFA of fish. It can be seen that the salmonids, even in freshwater, tend to have a higher total PUFA of the 20 and 22 carbon chain length, and a lower w 6/w 3 ratio than the other fish. The salmonids are mostly cold-water fish. The fatty acids from a number of marine animals from temperate and arctic waters show some significant differences in the general pattern; unfortunately analysis included fatty acids longer than 20:1. There are a number of other experiments demonstrating the effect of environmental temperature on fatty acid composition of aquatic animals. The general trend towards higher content of long chain PUFA at lower temperatures is quite clear. The w 6/w 3 ratio decreases with a decrease in temperature. If the trends in fatty acid composition can be taken as clues to the EFA requirements of fish, the w 3 requirement would be greater for fish raised at lower temperatures. Fish raised in warmer waters, such as common carp, channel catfish, and tilapia may do better with a mixture of w 6 and w 3 fatty acids.

Effects of Diet

Some of the fatty acid compositions may be seriously affected by the dietary lipids. The mosquito fish and guppies were fed trout pellets which had an w 6/ w 3 ratio of 2.75. The catfish were fed diets supplemented with either beef tallow or menhaden oil, with w 6/w 3 ratios of 18.13 and 0.15, respectively. These fish were able to alter the dietary w 6/w 3 ratio in favour of w 3 fatty acid incorporation into the flesh lipids even at the highest temperature. Commercially available trout pellets are often low in w 3 PUFA and high in w 6 fatty acids. It is important not to ignore the effect of dietary lipid composition on fatty acid composition of fish fed artificial diets. When the dietary ratio is very high in w 6 fatty acids supplied by animal lard or vegetable oils, there is a tendency for fish to alter the ratio of PUFA incorporated in favour of w 3 fatty acids. When the dietary oil is a fish oil high in to3 fatty acids, there is little change in the w 6/w 3 ratio of lipids incorporated into the fish. This is further suggestive evidence of an EFA requirement of fish for w 3 PUFA.

Seasonal Variation

Seasonal variations in the fatty acid composition of fish species have often been reported. Seasonal changes have been observed in total lipid and iodine values of herring oils. The iodine value or degree of unsaturation of the oil was minimal in April and maximal in June. The great increase in unsaturation corresponded to the onset of feeding in spring. The absence of a gas liquid chromatograph at the time precluded identification of changes in individual fatty acids.

Flesh and viscera lipid content of the sardine Sardinops melanosticta vary from 3.9 to 10.77 per cent and from 10.9 to 38.3 per cent, respectively. The fatty acids of principal interest with respect to EFA metabolism are 20:4w 6, 20:5w 3, and 22:6w 3. There was considerable variation in all of these fatty acids in both neutral and polar lipid from both tissues. In the flesh, the 20:4 w 6 was consistently higher in the neutral lipid than in the polar lipid.

The total 20:5w 3 plus 22:6w 3 was consistently higher in polar lipid than in the neutral lipid. Thus, in spite of the major fluctuations in fatty acids caused by changes in diet and temperature throughout the seasons, there was a consistent preferential incorporation of PUFA of the w 3 series into the polar or phospholipid fraction of the lipids.

One of the best clues to the EFA requirements of a species can be gained from the fatty acid composition of the lipids incorporated into the offspring or egg. The act of reproduction or spawning also has a significant effect on the seasonal fluctuation of lipids in fish. Fatty acid composition of fish egg lipids is probably distinctive for each species and contains increased levels of 16:0, 20:4 w 6, 20:5 w 3 and 22:6w 3 compared to the liver lipids of the same female fish.

Elevated levels of 16:0, 20:5w 3, and 22:6 w 3 and reduced 18:1 in the ovary occurred compared to mesenteric fat of Pacific sardine fed a natural copepod diet. The blood fatty acids of the sardine fed the natural diet were similar to those of the ovary. When the sardines were fed trout food, both the blood and mesenteric fat responded to the diet with elevated 18:2w 6 and reduced 20:5w 3 arid 22:6w 3. The effect of the diet on ovary fatty acid content was considerably less, as relatively high levels of 20:5w 3 and 22:6w 3 were retained.

The ovary lipids of the sweet smelt show an increase in 16:0, and a reduction in the PUFA, especially in the phospholipids, compared to the lipids from the flesh of fish caught at the same time of year. The w 6/w 3 ratio of the ovary was lower than that of the flesh lipids, 0.21 and 0.17 for the ovary compared to 0.31 and 0.20 for the triglycerids and phospholipids of the flesh, respectively.

The hatchability of eggs from common carp fed several different formulated feeds is greatly reduced when the 22:6w 3 of the egg lipids is less than 10 per cent. Further, the muscle, plasma, and erythrocyte fatty acid compositions are more affected by dietary lipid than those of the eggs.

The EFA requirements of a number of species of fish have been investigated in nutritional studies. The fish themselves have given ample evidence for EFA preference by the types of fatty acids they incorporate into their lipids. Fish, in general, tend to utilize w 3 over w 6.

This is especially observed when the dietary lipids are high in w 6, as the fish tend to alter the w 6/w 3 ratio towards the w 3 fatty acids in the tissue lipids. The lipids of the egg must satisfy the EFA requirement of the embryo until it is able to feed. The fatty acid composition data suggest that the w 3 requirement is greater in seawater than in freshwater and higher in cold water than in warm water.

AMINO ACIDS AND PROTEINS

AMINO ACIDS

The amino acids are the building blocks of proteins; about 23 amino acids have been isolated from natural proteins. Ten of these are indispensable for fish. The animal is incapable of synthesizing indispensable amino acids and must therefore obtain these from the diet.

Essential and Non-essential

Salmon, trout and channel catfish fed diets devoid of arginine, histidine, isoleucine, leucine, lysine, methionine, phenylalanine, threonine, tryptophan or valine failed to grow.

These same fish fed diets devoid of other L-amino acids grew as well as fish receiving all 18 amino acids tested.

The nitrogen component in the test diets was made up of 18 L-amino acids in the pattern found in whole egg protein. All fish on test recovered rapidly when the missing amino acid was replaced in the diet. The slope of the growth curve of the recovery group was identical with that of fish receiving the complete amino acid test diet.

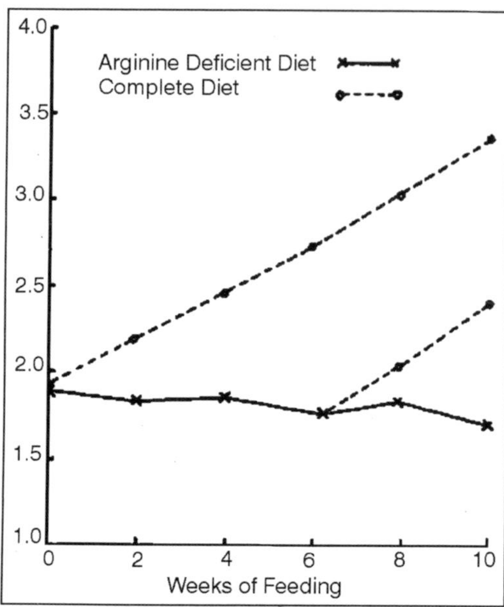

Fig. Growth of Cystine Deficient

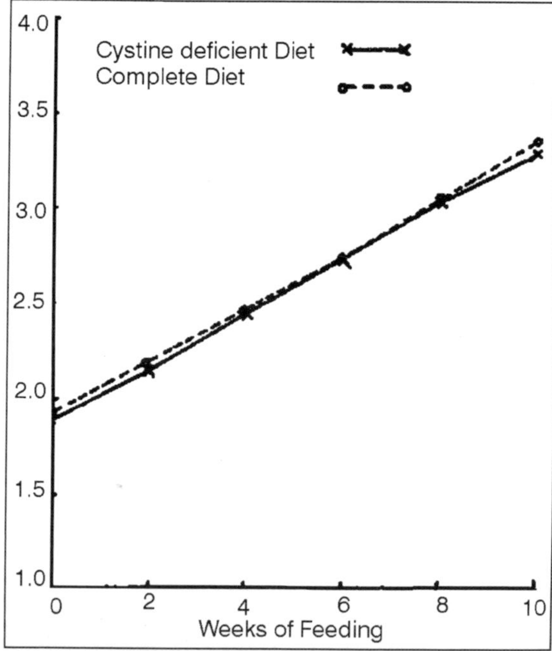

Dispensable amino acids tested were alanine, aspartic acid, cystine, glutamic acid, glycine, proline, serine, and tyrosine. These amino acids were found to be not essential for the growth of salmon, trout and channel catfish.

Quantitative studies on the requirements of the 10 indispensable amino acids used a casein-gelatin mixture supplemented with crystalline L-amino acids. The test diet had an amino acid pattern of 40 per cent whole egg protein for the nitrogen component.

Experiments conducted with carp and eel showed a similar lack of growth when an indispensable amino acid was absent from the diet.

Essential Amino Acids and Protein Quality

If the essential amino acid requirements of fish are known, it should be possible to meet these needs in culture systems in a number of ways from different food proteins or combinations of food proteins.

Phenylalanine is spared by tyrosine. It is not known to be chemically modified nor rendered unavailable by the harsh conditions to which feedstuff proteins are normally subjected during processing. Measurement of phenylalanine in proteins is uncomplicated so that the provision and evaluation of phenylalanine in proteins in practical diets presents little difficulty.

Lysine is a basic amino acid. In addition to the -amino acid group normally bound in peptide linkage, it also contains a second, -amino group. This -amino group must be free and reactive, otherwise the lysine, although chemically measurable, will not be biologically available. During the processing of feedstuff proteins the -amino group of lysine may react with non-protein molecules present in the feedstuff to form additional compounds that render the lysine biologically unavailable.

Methionine is spared by cystine. However, measurement of the methionine content of feed proteins is not easy as the amino acid is subject to oxidation during processing. After processing, methionine may be present as such or as the sulphoxide or as the sulphone. The sulphoxide may be formed from methionine during acid hydrolysis of the feed protein prior to measurement of its any-no acid composition.

Acid hydrolysis of proteins before analysis disturbs the original equilibrium between the two compounds so that the composition of the hydrolysate no longer reflects that of the protein. In determining the methionine content of pure proteins, oxidation of the amino acid to methionine sulphone is normally quantitative. In the case of feed proteins, however, this will not reveal how much methionine or methionine sulphoxide was present in the protein prior to performate oxidation and hydrolysis. Methionine sulphoxide may have some biological value for fish which may have some capability of reconverting it to methionine and thus partially make up for some of the methionine oxidized during processing.

Methods have recently been reported for measurement of methionine in proteins using an iodoplatinate reagent before and after reduction with titanium trichloride, to give values for both methionine and the sulphoxide in the original protein. A method for measuring methionine specifically by cyanogen bromide cleavage has also been described. Both methods remain to be independently assessed. Microbiological assay of methionine in feed proteins is a valuable tool although there is the danger that oxides of methionine may differ in their activity for micro-organisms and misrepresent values.

QUANTITATIVE REQUIREMENTS

Quantitative requirements by salmonids for the ten indispensable amino acids were determined by feeding linear increments of one amino acid at a time in a test diet containing an amino acid profile identical with whole egg protein except for the amino acid tested. Replicate groups of fish were fed the diet treatments until gross differences appeared in the growth of test lots.

An Almquist plot of growth response indicated the level of amino acids required for maximum growth under those specific test conditions. Diets were designed to contain protein at or slightly below the optimum protein requirement for that species and test condition to assure maximum utilization of the limiting amino acid. A comparison of the requirements for the ten indispensable amino acids between species.

A recent innovation has been the use in test diets of proteins relatively deficient in a given essential amino acid. Thus combinations of fishmeal and zein have been used in test diets to define the requirement of rainbow trout for arginine. Diets containing different relative amounts of casein and gelatin showed that an increase in the level of protein-bound arginine from 11 to 17 g/kg resulted in a significant increase in the growth of channel catfish.

Table. Amino Acid Requirements of Seven Animals

Amino acid	Eel fingerling	Carp fry	Channel catfish	Chinook salmon fingerling	Chick	Young Pig	Rat
Arginine	3.9 (1.7/42)	4.3 (1.65/38.5)		6.0 (2.4/40)	6.1 (1.1/18)	1.5 (0.2/13)	1.0 (0.2/19)
Histidine	1.9 (0.8/42)			1.8 (0.7/40)	1.7 (0.3/18)	1.5 (0.2/13)	2.1 (0.4/19)
Isoleucine	3.6 (1.5/42)	2.6 (1.0/38.5)		2.2 (0.9/41)	4.4 (0.8/18)	4.6 (0.6/13)	3.9 (0.5/13)
Leucine	4.1 (1.7/42)	3.9 (1.5/38.5)		3.9 (1.6/41)	6.7 (1.2/18)	4.6 (0.6/13)	4.5 (0.9/19)
Lysine	4.8 (2.0/42)		5.1 (1.23/24.0)	5.0 (2.0/40)	6.1 (1.1/18)	4.7 (0.65/13)	5.4 (1.0/19)
Methionine	4.5 (2.1/42)	3.1 (1.2/38.5)	2.3 (0.56/24.0)	4.0 (1.6/40)	4.4 (0.8/18)	3.0 (0.6/20)	3.0 (0.6/20)
Phenylalanine				5.1 (2.1/41)	7.2 (1.3/18)	3.6 (0.45/13)	5.3 (0.9/17)
Threonine	3.6 (1.5/42)			2.2 (0.9/40)	3.3 (0.6/18)	3.0 (0.4/13)	3.1 (0.2/19)
Tryptophan	1.0 (0.4/42)			0.5 (0.2/40)	1.1 (0.2/18)	0.8 (0.2/25)	1.0 (0.2/19)
Valine	3.6 (1.5/42)			3.2 (1.3/40)	4.4 (0.8/18)	3.1 (0.4/13)	3.1 (0.4/13)

Arginine requirement of rainbow trout has been determined from a conventional dose/response curve and also by measuring the blood and muscle levels of free arginine in groups of trout given increasing amounts of dietary arginine. After the dietary requirement of the trout for arginine has been met, any further increase in arginine intake led to an increase in the concentration of free arginine in blood and muscle. Good agreement was obtained between the two methods. That real differences exist between fish species in their requirement for certain amino acids. This leads to difficulties in formulating the protein component of practical diets for those species whose amino acid requirements are not yet known. A possible solution is to use, for each amino acid, the highest level required by any of those species for which data is available. The need for further quantitative data on the amino acid requirements of fish, especially those actually or potentially useful as farm animals, is obvious.

Supplementing Diets

One solution to the use of proteins that are relatively deficient in one or more amino acids is to supplement the protein with appropriate amounts of the amino acid needed in practical diets. Fish appear to utilize free amino acids at various degrees of efficiency.

Young carp, Cyprinus carpio, were shown to be unable to grow on diets in which the protein component was replaced by a mixture of amino acids similar in overall composition.

A trypsin hydrolyzate of casein was equally ineffective. However, if a diet containing free amino acids as the protein component is carefully neutralized with NaOH to pH 6.5-6.7 then some growth of young carp does occur. This growth was markedly inferior to that occurring on a comparable casein diet under the same conditions.

Channel catfish are also unable to utilize free amino acids given as supplements to deficient proteins. When soybean meal was substituted isonitrogenously for menhaden meal, growth and feed efficiency of channel catfish were substantially reduced. Addition of free methionine, cystine or lysine, the most limiting amino acids, to these soy-substituted diets did not enhance weight gain. Raising the arginine level of catfish diets from 11 to 17 g/kg by isonitrogenous substitution of gelatin for casein enhanced weight gain significantly but the addition of free arginine, cystine, tryptophan or methionine to casein had little effect on growth or food conversion.

Salmonids are able to utilize free amino acids for growth. A zein-gelatin diet supplemented with lysine and trytophan was shown to be markedly superior to an unsupplemented zein-gelatin diet for rainbow trout when weight gain and protein utilization were used as criteria.

Several investigators have demonstrated the potential of supplementing amino acid deficient proteins with limiting amino acids in diets for salmonids. Casein supplemented with six amino acids produced feed conversion ratios with Atlantic salmon similar to those obtained when an isolated fish protein was used as the dietary protein source. Soybean meal supplemented with five or more amino acids was a superior protein source to soybean meal alone for rainbow trout. Single additions of methionine and lysine did not, however, improve the value of soybean meal. These results suggest that the amino acid spectrum of the isolated fish protein they used may possibly approximate the amino acid requirement of rainbow trout. The nutritional value of a soy protein isolate could be enhanced by supplementing it with the first limiting amino acid; *i.e.*, methionine. Diets containing, as protein component, fishmeal, meat and bone meal, and yeast and soybean meal could be improved by supplementing with cystine and tryptophan together. Fishmeal can be entirely replaced without a reduction in food conversion rate in diets for rainbow trout by a mixture of poultry by-product meal and feather meal together with 17 g lysine HCL/kg, 4.8 g DL-methionine/kg, and 1.44 g DL-tryptophan/kg.

PROTEINS

Proteins are complex, organic compounds composed of many amino acids linked together through peptide bonds and cross-linked between chains by sulfhydryl bonds, hydrogen bonds and van der Waals forces. There is a greater diversity of chemical composition in proteins than in any other group of biologically active compounds. The proteins in the various animal and plant cells confer on these tissues their biological specificity.

Classification

Proteins can be classified as:

- *Simple Proteins*: On hydrolysis they yield only the amino acids and occasional small carbohydrate compounds. Examples are: albumins, globulins, glutelins, albuminoids, histones and protamines.
- *Conjugated Proteins*: These are simple proteins combined with some non-protein material in the body. Examples are: nucleoproteins, glycoproteins, phosphoproteins, haemoglobins and lecithoproteins.
- *Derived Proteins*: These are proteins derived from simple or conjugated proteins by physical or chemical means. Examples are: denatured proteins and peptides.

Structure

The potential configuration of protein molecules is so complex that many types of protein molecules can be constructed and are found in biological materials with different physical characteristics. Globular proteins are found in blood and tissue fluids in amorphous globular form with very thin or non-existent membranes. Collagenous proteins are found in connective tissue such as skin or cell membranes. Fibrous proteins are found in hair, muscle and connective tissue. Crystalline proteins are exemplified by the lens of the eye and similar tissues.

Enzymes are proteins with specific chemical functions and mediate most of the physiological processes of life. Several small polypeptides act as hormones in tissue systems controlling different chemical or physiological processes. Muscle protein is made of several forms of polypeptides that allow muscular contraction and relaxation for physical movement.

Properties

Proteins can also be characterized by their chemical reactions. Most proteins are soluble in water, in alcohol, in dilute base or in various concentrations of salt solutions. Proteins have the characteristic coiled structure which is determined by the sequence of amino acids in the primary polypeptide chain and the stereo configuration of the radical groups attached to the alpha carbon of each amino acid.

Proteins are heat labile exhibiting various degrees of lability depending upon type of protein, solution and temperature profile. Proteins can be reversible or irreversible, denatured by heating, by salt concentration, by freezing, by ultrasonic stress or by aging. Proteins undergo characteristic bonding with other proteins in the so-called plastein reaction and will combine with free aldyhyde and hydroxy groups of carbohydrates to form Maillard type compounds.

Chemical Determination

The nitrogen content of most proteins found in animal, nut and grain tissue is about 16 per cent; therefore, protein content is commonly expressed as nitrogen content × 6.25.

PROTEIN DIGESTION AND METABOLISM

Ingested proteins are first split into smaller fragments by pepsin in the stomach or by trypsin or chymotrypsin from the pancreas. These peptides are then further reduced by the action of carboxypeptidase which hydrolyzes off one amino acid at a time beginning at the free carboxyl end of the molecule or by aminopeptidase which splits off one amino acid at a time beginning at the free amino end of the polypeptide chain.

The free amino acids released into the digestive system are then absorbed through the walls of the gastro intestinal tract into the blood stream where they are then resynthesized into new tissue proteins or are catabolyzed for energy or for fragments for further tissue metabolism.

GROSS PROTEIN REQUIREMENTS

Gross protein requirements have been determined for a few species of fish. Simulated whole egg protein component of test diets contains an excess of indispensable amino acids. These diets were kept approximately isocaloric by adjusting total protein plus digestible carbohydrate components to a fixed amount as the protein diet treatments were varied over the ranges tested.

Tests in feeding fry, fingerling, and yearling fish have shown that gross protein requirements are highest in initial feeding fry and that they decrease as fish size increases. To grow at the maximum rate, fry must have a diet in which nearly half of the digestible ingredients consist of balanced protein; at 6-8 weeks this requirement is decreased to about 40 per cent of the diet for salmon and trout and to about 35 per cent of the diet for yearling salmonids raised at standard environmental temperature (SET).

Gross protein requirements for young Catfish appear to be less than those for salmonids. Initially feeding fry require that about 50 per cent of the digestible components of the ration be protein, and the requirement decreases with size. Some feeding trials with salmon have indicated direct relationships between changes in the protein requirements of young fish and changes in water temperature.

Chinook salmon in 7o C water require about 40 per cent whole egg protein for maximum growth; the same fish in 15 o C water require about 50 per cent protein. Salmon, trout and catfish can use more protein than required for maximum growth because of efficiency in eliminating nitrogenous wastes in the form of soluble ammonia compounds through the gill tissue directly into the water environment. This system for eliminating nitrogen is more efficient than that available to fowl and mammals. Fowl and mammals consume energy to synthesize urea, uric acid, or other nitrogen compounds which are excreted through the kidney tissue and expelled in urine. Digestible carbohydrate and fat will spare excess protein in the diet as long as the protein requirement for maximum growth is met.

Table. Estimated Dietary Protein Requirement

Species	Crude protein level in diet for optimal growth (g/kg)
Rainbow trout	400-460
Carp	380
Chinook salmon	400
Eel	445
Plaice	500
Gilthead bream	400
Grass carp	410-430
Brycon sp.	356
Red sea bream	550
Yellowtail	550

Basically the fish must be given a diet containing graded levels of high quality protein and energy and adequate balances of essential fatty acids, vitamins and minerals over a prolonged period. From the resulting dose/response curve the protein requirement is usually obtained by an Almquist plot. These differences in apparent protein requirement are thought to be due to differences in culture techniques and diet composition.

The relatively high dietary protein levels required for maximal growth of certain fish such as grass carp, Ctenopharyngodon idella, and Brycon spp. are surprising as these fish are omnivorous. Brycon spp. are grown on unwanted fruit and other plant material of low protein content and under these conditions there is presumably a substantial contribution to their protein intake from a natural food chain. Protein requirement of eurythaline fish such as the rainbow trout, Salmo gairdneri, and the coho salmon, Oncorhynchus kisutch, reared in water of salinity 20 ppt is about the same as the requirement in freshwater. No data are available for the protein requirement of these species in full strength sea water.

LAND-USE ISSUES IN AQUACULTURE

Regions suitable for land-based aquaculture are limited by the right combination of climate, soil type, geomorphology, water availability, and access to other resources such as labour, markets, feeds, and larvae or juveniles for stocking. Nevertheless, land use per se is seldom a limiting factor in aquaculture development. This is particularly true in the United States, where relatively few locations are significantly developed into aquaculture. The Yazoo-Mississippi River alluvial valley in northwest Mississippi is the most intensely developed aquaculture region in the United States, with aquaculture ponds occupying more than 45,000 ha of land. Yet ponds account for only about 2 per cent of land use in the region. Even in sub-watersheds within the Yazoo basin with more intense development, ponds account for only 10 to 20 per cent of the watershed. Aquaculture development in the United States is more likely constrained by water availability, regulatory constraints on land development, or the value of land for alternative uses than by the availability of suitable land for facility construction.

At times, availability of the right combination of resources leads to localized overdevelopment of aquaculture. In this respect the problem is not necessarily the physical concentration of facilities, but rather ecological imbalances associated with the inability of the local ecosystem to provide support services, such as adequate water supply or waste treatment.

A good example is the study of the ecosystem support area needed for semi-intensive shrimp farming in the Bay of Barbacoas area of Colombia, South America. The estimated local ecosystem support area needed to assimilate waste nutrients from shrimp farms closely matched the mangrove area available to provide that service. Local capacity for waste treatment was used to its fullest extent and additional development could lead to self-pollution of the water supply, spread of disease, and other environmental impacts that have been associated with the collapse of shrimp farming in other overdeveloped areas.

Most problems associated with land use in aquaculture are related to the value of alternative uses of the land. These conflicts arise when aquaculture must compete with other land uses. Land-use conflicts tend to be especially acute in coastal regions, which are limited in area, have proportionately large areas of sensitive and environmentally valuable ecosystems, and are intensively used for a range of often conflicting purposes by dense human populations. Coastal regions also tend to be more commercially developed and the land more costly, especially in the United States. Clearing mangrove forests to build ponds for shrimp and fish culture has been one of the most contentious land use conflicts in aquaculture. Mangrove ecosystems have high inherent aesthetic and environmental value.

They protect coastlines from erosion and flooding and provide habitat and food for a tremendous variety of aquatic, terrestrial, and avian species. Mangroves also act as natural purification systems for waters interacting at the interface of freshwater and marine environments. They are also convenient—although far from ideal—locations for brackishwa-ter aquaculture ponds, especially for growing shrimp. In the 1980s and early 1990s, when global shrimp aquaculture was expanding rapidly, thousands of square kilometers of ponds were built in mangrove areas. Much of this land had been cleared previously for other uses, but the destruction was often attributed to shrimp farming. Although environmental advocacy groups overstated the involvement of shrimp farming in the historic loss of mangrove, worldwide awareness of the problem resulted in stricter governmental oversight of the use of mangrove forests and stimulated significant changes in shrimp production practices to protect these valuable ecosystems.

Aquaculture can affect sensitive ecosystems other than mangroves. Marine benthic communities may be impacted by net-pen aquaculture, and construction and operation of freshwater ponds can destroy sensitive wetlands. Certain types of open-water molluscan shellfish culture—usually viewed as an environmentally benign endeavor—can damage seagrass beds due to siltation and the mechanical effects of certain culture practices.

Marine aquaculture also faces unique conflicts unrelated to the physical area occupied by the facility or the impacts of culture activities. Conflicts may arise from competition with other users of the area or region, including activities such as tourism, shipping and navigation, capture fisheries, and urban development. Conflicts may even arise from esthetic considerations, including the effect of aquaculture development on the scenic quality of an area (the *viewscape*) and disturbances related to increased activity and noise associated with culture practices.

Nearly all problems arising from land-use conflicts in aquaculture can be avoided by prudent site selection. Better management practices for facility site selection.

AGRICULTURE AND WATER RESOURCES

Cultivated systems, including freshwater aquaculture, now cover 24 per cent of the terrestrial surface of the planet. Most of the increase in food production in the last 50 years has been derived from agricultural intensification, not an increase in cropland area. Intensification has increased pressure on water resources (both for supply and waste treatment) and fisheries and cropland resources for feed, and it requires increased energy resources, nearly all derived from fossil fuels. Freshwater is a scarce resource on the planet, with less than 1 per cent of total freshwater resources available for human use.

Renewable freshwater resources available for human use are derived from precipitation that falls on land. This precipitation runs off (including infiltration to groundwater) or is lost by evapotranspiration from land and water surfaces and through plants. Humans use about 10 per cent of renewable freshwater supplies, although some countries use more than 100 per cent of renewable supplies.

About 40 to 50 per cent of freshwater run-off is appropriated for human use. The available supplies of global freshwater are used by agriculture (75 per cent), by industry (20 per cent), and for domestic purposes (5 per cent).

Aquatic resources are impacted by numerous human activities. Running waters are exploited to supply irrigation and drinking water, generate electricity, and receive wastes. Running waters are also affected by land use in adjacent watersheds.

Threats to running waters include alteration of habitat; changes in water chemistry; and changes in biodiversity caused by land transformations, channelization, and industrial or urban development. Standing waters provide water for agricultural, industrial, and domestic uses. Threats to standing water ecosystems include eutrophication, chemical contamination, and the introduction of invasive species.

WATER USE IN AQUACULTURE

Water is an obvious and essential resource for aquaculture. No single factor influences the success of aquaculture more than availability of an adequate supply of good quality water. Availability of freshwater increasingly constrains all forms of human development, so aquaculture must compete with other uses, including irrigated agriculture, industry, and urbanization. Competition for finite water supplies is particularly acute in arid regions or where freshwater is supplied primarily from groundwater resources. Sparing water for other uses by adopting conservative water use in aquaculture is therefore a social obligation. Water use is also inextricably tied to issues of waste discharge because water input and discharge are variables on two sides of the hydrological equation: increasing water input volume will increase water volume discharged.

Water use in aquaculture may be classified as either total use or consumptive use. Total water use is the sum of all inflows (precipitation, run-off, seepage inflow, and management additions) to production facilities. Depending on the type of culture system, a portion of the water entering the facility passes downstream in overflow or intentional discharge and is available for other purposes (*e.g.*, ecosystem support, irrigation). For example, essentially all water flowing into flow-through systems is discharged, whereas some ponds have no discharge at all. Consumptive use can be defined as water used in aquaculture that is not available for other purposes. Consumptive water use includes water lost by evaporation and seepage from an aquaculture facility, withdrawal of fresh ground-water, and water removed in biomass of aquatic animals at harvest. Water in harvest biomass averages about 0.75 L/kg, a minor quantity compared to other uses.

POTENTIAL ENVIRONMENTAL IMPACTS OF AQUACULTURE

The environmental effects of aquaculture are often described in terms of the sources (type and quantity) or causes of a potential effect, and this convention will be followed here. Impacts are potential because effects depend on characteristics of the environment in which aquaculture production systems are embedded. For example, releasing waste into a closed basin will have a qualitatively greater (more adverse) impact that releasing the same quantity and type of waste into an open basin.

Contrasting to broad issues related to resource use, many of the problems commonly attributed to aquaculture—such as the adverse effects of habitat conversion, waste discharge, consumptive water use, and release of chemicals and antibiotics—are related to facility operations and can be addressed on shorter time scales through technology-based approaches. Solutions to problems of this nature are the focus of this book.

HABITAT CONVERSION

Ponds are the most common aquaculture production system in the United States and the world. In most cases, pond construction requires conversion of terrestrial or wetland habitat to aquatic habitat. Aquaculture ponds must be located near sources of water, so they are commonly constructed in coastal or riverine floodplains. Thus, pond construction may convert floodplain land or wetlands to ponds.

Pond construction in or near coastal wetlands, particularly the conversion of tropical mangroves to shrimp ponds, is controversial. Coastal wetlands are multiple-use, mostly open-access resources that are increasingly threatened by a wide range of development pressures, including aquaculture. Coastal wetlands provide a number of ecosystem services, including waste treatment, storm protection, food production (especially as a nursery for many aquatic species), recreation, and a source of raw materials.

The environmental impact of pond construction in coastal wetlands can be described in terms of the degree to which these services are lost by conversion to ponds. The total economic value (including the value of non-marketed ecosystem services) of sustainably managed mangrove wetlands often exceeds the value associated with conversion to shrimp ponds. Coastal development, including aquaculture, in tropical areas can adversely affect adjacent, light-sensitive ecosystems, such as coral reefs and seagrass meadows. Approximately 35 per cent of mangroves have been lost in the last 2 decades. Estimates of the relative contribution of aquaculture to the loss of coastal mangroves are uncertain, although less than 10 per cent of the global loss of mangrove area has been attributed to conversion into shrimp ponds. Nonetheless, conversion to aquaculture ponds can be the major component of mangrove loss in some coastal areas, particularly in southeast Asia. Mangrove wetlands are increasingly seen as having intrinsic value, being poor sites for pond construction, and undesirable sources of social conflict when aquaculture projects are developed. Pond aquaculture in coastal wetlands is much less common than in the past. In many places in Asia, mangrove clearing has been banned and rehabilitation and restoration programmes are active.

Waste Loading

Nearly all aquaculture production systems produce waste nutrients and organic matter. Depending on the production method, organic fertilizers (manures and agricultural byproducts), inorganic fertilizers, and nutrient-dense feeds are applied to production units in most aquaculture facilities to promote growth of finfish, crustaceans, and other species. Approximately 70 to 80 per cent of nutrients in feed are released as waste into the culture unit in aquaculture production systems where cultured animals are fed.

The degree to which the culture unit is open to the environment will determine the proportion of waste excreted by cultured animals that is released to the environment. For example, ponds and recirculating systems are operated with a long hydraulic residence time, so a very small proportion of nutrients are released to the environment. In contrast, culture systems with a more direct hydrological connection with the outside environment, such as flow-through systems and net pens, will release a large proportion of waste nutrients and organic matter.

The ecological effects of discharged nutrients will vary depending on production intensity, facility density, the hydraulic retention time and trophic state of receiving waters, and dilution rate. In relatively closed production systems such as ponds, feeding invariably results in the eutrophication of pond water. In more open systems, the effects of nutrient discharge are far less predictable. If the flushing time of the water body is less than the generation time of phytoplankton, an increase in limiting nutrient concentration can result in eutrophication. It is important to emphasize the point that phytoplankton respond to changes in nutrient concentration, particularly of those nutrients that tend to limit production (phosphorus in freshwater and nitrogen in seawater), not to increases in nutrient mass loading per se.

Potential adverse environmental effects attributed to organic waste loading from aqua-culture include increases in phytoplankton density in response to elevated nutrient concentration, increased variation in dissolved oxygen concentration in response to increased phytoplankton density, increased frequency of dissolved oxygen depletion, reduced dissolved oxygen concentration caused by the discharge of effluent with high biochemical oxygen demand, and changes to benthic communities caused by localized sedimentation of suspended solids. Nutrients discharged from aquaculture facilities have not been linked to the development of harmful algal blooms.

In many coastal areas, discharges from aquaculture production facilities represent one of many nutrient sources contributing to eutrophication. Truly addressing problems related to eutrophication therefore requires that the relative contribution of aquaculture be evaluated in terms of all sources and considered in the context of a watershed or whole-basin approach. In certain locations, particularly closed basins, aqua-culture operations can be the primary contributor to localized eutrophication. The effects of effluent discharge tend to be localized near the discharge point. The susceptibility of receiving waters to eutrophication depends on trophic status prior to enrichment. For a given level of nutrient loading, oligotrophic waters are more susceptible to trophic state changes than mesotrophic waters.

Most discussions of waste loading from aquaculture consider only the components of the waste stream, which includes nutrients, organic matter, potential pathogens, chemicals or therapeutants, and escaped fish. However, they do not consider how the environment responds to waste loading. Often missing is the notion of environmental carrying capacity, which requires an understanding of the waste assimilation capacity of the environment and consensus on an acceptable level of environmental change.

Aquatic environments have an inherent capacity to assimilate nutrients through well-known biological and physicochemical processes. In some areas, aquaculture facilities are sufficiently concentrated and production intensity is sufficiently great that the majority of the assimilative capacity of receiving waters is used by aquaculture. Appropriation of the waste treatment ecosystem service by aquaculture has been criticized on the basis that the cost of waste treatment is externalized by aquaculture operations and born by society. In general, nutrient enrichment is perceived to have negative consequences, although in some environments, eutrophication may be beneficial and perceived as desirable. Changes in trophic state have no inherent value; the acceptability of the change is a matter of societal values and policy.

Benthic Impacts

Organic wastes produced in open aquaculture production systems can accumulate on the sediment in the vicinity of culture units. Soil eroded from pond embankments and organic matter in pond water can settle in receiving water bodies adjacent to discharge points. Organic matter discharged from flow-through systems can accumulate near the discharge point, leading to localized suppression of dissolved oxygen concentration and changes in benthic communities. Depending on current speed and water depth, organic matter mainly derived from feces and uneaten feed can accumulate on the sea floor beneath net pens. Sedimentation rates within net-pen facilities are reported to range from about 15 to 100 g of total volatile solids (TVS)/m^2 per day, with most estimates between 25 and 50 g TVS/m^2 per day. The accumulation of organic matter can extend from 145 to 205 m downcurrent from the perimeter of net-pen facilities, although significant effects are usually restricted to less than 60 m from the perimeter. Of all the potential environmental impacts of salmonid net-pen aquaculture, changes to the sediment beneath net pens are considered to represent the greatest risk to the environment, although these changes are temporary and largely reversible.

If the rate of organic matter sedimentation exceeds the rate of decomposition, organic matter will accumulate. Organic matter accumulation will depend on loading rate, temperature, oxygen supply, current velocity, and other physical and chemical factors. The deposition of organic matter creates a dissolved oxygen demand for decomposition that may exceed supply. In this case, the redox potential will shift to lower (more negative) values, indicating anaerobic conditions. Mats of the sulfide-oxidizing filamentous bacteria *Beggiatoa* may cover the sediment surface. Reduced substances such as ammonia, dissolved phosphorus, methane, carbon dioxide, and especially in marine waters, hydrogen sulfide will diffuse into the overlying water. Some of these substances are potentially toxic to benthic invertebrates and cultured fish, but rarely accumulate to levels that will affect fish culture performance.

Anaerobic conditions in the sediment will also cause a shift in the assemblage of benthic invertebrates towards a less diverse community dominated by pollution-tolerant species. The effects described here tend to be localized around effluent discharge points and within 25 m of the perimeter of net-pen farms (Karakassis *et al.* 2000; Pearson and Black 2001). These effects are also more pronounced for poorly flushed sites (current velocity <10 cm/second) with fine-grained sediments. In well-flushed sites (>50 cm/second), the abundance and diversity of benthic infauna frequently increases. Chemical remediation of marine sediments beneath net pens occurs within months to 1 year, but biological remediation may require 2 to 3 years in some sites, largely depending on water temperature, current velocity, and concentration of sediment organic matter. In general, the effects of sedimentation are localized and reversible.

Sedimentation can also occur as a result of alteration in hydrodynamics caused by the deployment of aquaculture structures such as rafts, longlines, and cages. These can reduce current velocity in the lee of culture facilities, creating a depositional environment. In addition, feces and psuedofeces produced by shellfish can accumulate in the vicinity of shellfish rafts. Hydrodynamics can be altered by material that accumulates near effluent outfalls, which can reduce stream depth and obstruct water flows, increasing the potential for flooding. Some metals will accumulate in sediment, but there is no evidence that these accumulations have had negative impacts on benthic invertebrates or rates of biogeo-chemical transformations. Zinc is added to fish feeds as an essential nutrient, and elevated concentrations have been measured in sediment beneath net pens. Copper is used in antifouling paints and net treatments and can accumulate in sediment beneath net pens. In marine environments, the accumulation and bioavailability of copper and zinc depend on sulphide concentration, with reduced bioavailability with elevated sulphide concentration. Concentrations decline during site fallowing.

Chemical Pollution

As with other forms of agriculture, chemicals are used in aquaculture for a broad diversity of purposes. Perhaps most importantly, pesticides, disinfectants, and antibiotics are used for disease treatment and pest management. Chemicals are also used for soil and water treatment, enhancing natural productivity, feed manufacture, control of reproduction, growth enhancement, transportation, and processing. Major classes of chemicals include therapeutants (including antibacterial agents), disinfectants, anesthetics, hormones, feed additives, antifouling paints, herbicides, and pesticides. Soil and water are treated with chemicals such as inorganic and organic fertilizers, liming agents, flocculating agents, and bacterial amendments. Chemical and drug use are highly regulated in the United States, and even usage of legally registered substances is relatively uncommon for reasons of economics. This is not necessarily true elsewhere, however. On average, a shrimp farmer in Thailand uses 13 chemicals, with four pesticides and disinfectants, and three soil and water treatment products. Thai shrimp farmers use an average of one antibiotic per farm, most commonly fluoroquinolones.

Chemicals used in aquaculture can be divided into three groups based on the different ways they affect the environment. One group of chemicals encompasses those with acute or chronic toxicity to non-target organisms, including cultured animals. Another group are the antibiotics that are released to the environment. These can lead to the development of resistant strains of pathogenic bacteria and alter microbial community composition, affecting biogeochemical processes.

The third group are nutrients that can lead to eutrophication if discharged, a particular concern for light-sensitive seagrass and coral reef ecosystems often found adjacent to coastal shrimp ponds. Chemicals can also be grouped into three categories based on the environmental or human health risk associated with their use. The first group includes chemicals with a high risk associated with their use, such as chloramphenicol, malachite green, and organophosphates. Chemicals in a second group can be used safely if applied according to label directions, but may present an environmental or human health hazard if used incorrectly. A third group of chemicals can be used safely at most locations, but may present an elevated risk at certain sites because of particular characteristics of that site.

Chemicals have various fates when released from aquaculture production systems to the environment, so environmental persistence is variable. The persistence of chemicals in the environment depends on physicochemical factors that affect solubility and reactivity (especially temperature, pH, dissolved oxygen concentration, and light intensity) and biological factors, especially the type and density of microorganisms. Some chemicals can accumulate in sediments. Other chemicals dissipate or degrade by physicochemical reactions such as photooxidation or adsorption to sediment minerals or organic matter. Others are transformed or degraded biologically by reactions mediated by microorganisms. The environmental toxicity of reaction products can be greater or less than that of reaction substrates. In general, but with some important exceptions, most chemicals used in aqua-culture do not persist in the environment.

The environmental and human health issues associated with chemical use in aquaculture were summarized by GESAMP. These include persistence in aquatic environments, residues in non-cultured organisms, toxicity to non-target organisms, stimulation of bacterial resistance, effects on sediment biogeochemistry, nutrient enrichment, health of workers exposed to chemicals, and residues in seafood that affect product safety. Many of these effects can disturb natural aquatic communities, altering community structure and reducing biodiversity.

There is a strong disincentive to use many chemicals in aquaculture because use increases production costs and some chemicals may be toxic to cultured aquatic animals. Chemical use also presents a food safety problem by increasing the risk of consumption of aquatic animals containing chemical residues. Improper use of antimicrobials can stimulate the development of drug-resistant forms of microbial pathogens in cultured animals, water, or sediment. Chemicals can also bioaccumulate in non-target organisms and become biomagnified as they move through aquatic food webs to higher trophic levels.

In the particular case of salmonid net-pen aquaculture, there has been concern about the use and fate of pesticides and drugs applied to control infestations of various species of parasitic copepods known collectively as sea lice and the antibiotics used to control bacterial diseases. In some countries, sea lice are controlled with bath treatments of non-specific and broad-spectrum organophosphorus and pyrethroid pesticides and with the antibiotic ivermectin incorporated into feed. The chemicals used in bath treatments disperse rapidly and are usually not detected beyond 25 m from a net pen, although these chemicals have the potential to affect non-target organisms, particularly crustaceans. Research on the fate of antibiotics used to control bacterial diseases indicates that antibiotic residues will accumulate and persist in sediment beneath cage sites and have been measured in non-target organisms. There is concern about the increased risk of the development of drug-resistant strains of bacteria that are pathogenic to fish, particularly when antibiotics are used prophylactically. Similar concerns have been expressed about the development of drug-resistant strains of bacteria in shrimp farming.

Salinization

Salinization refers to increases in the salinity of soil, surface water, and groundwater. Excessive salinization of water and soil can limit the use of surface waters for irrigation, reduce crop yields, displace salt-intolerant crops, and limit the use of water for domestic purposes. Salinization is primarily a concern with coastal pond aquaculture, where water is pumped to maintain appropriate salinity in ponds for penaeid shrimp culture. However, salinization has become a concern with the increase in inland aquaculture of marine shrimp. These facilities are supplied with brine transported from coastal waters or saline groundwater from wells, and salinization can occur when this water is discharged to inland streams. Salinization can also occur when accumulated sediment is removed from coastal aquaculture ponds and disposed in freshwater areas. Salt contained in the sediment leaches out following rainfall and can cause salinization.

The discharge of water from coastal or inland brackishwater ponds into freshwater bodies can impair surface water quality to limit their use, especially for agriculture. In sites with improperly compacted soils, water can seep from these ponds into underlying freshwater aquifers. Alternatively, excessive pumping of groundwater can result in saltwater intrusion of aquifers, thereby limiting use. Excessive pumping of groundwater to control salinity in coastal aquaculture ponds has led to salinization of groundwater aquifers and land subsidence and consequent damage to infrastructure in Taiwan. Pumping groundwater was widely used to manage salinity in brackishwater ponds in the past, but the practice is now banned or restricted in many places.

Pathogen Transmission

In aquaculture production systems, animals are usually held at much greater densities than in the natural environment, a condition that facilitates the transmission of pathogens among cultured animals. Elevated production intensity can lead to environmental conditions in the production unit that stress cultured animals, potentially increasing the susceptibility to infection by pathogens that may lead to disease. Thus, the conditions that promote disease are more prevalent in aquaculture production systems than in the natural environment. Intensive culture conditions can also cause amplification of pathogens that are ubiquitous in the environment.

Pathogens are a constituent of water that is discharged from or flows through aquacul-ture facilities. Water exchange in open production systems facilitates the transmission of pathogens to the environment, specifically between cultured and wild fish. There are numerous examples of diseases caused by the discharge of pathogens from aquaculture production facilities. Wild fish also serve as a reservoir of pathogens for the infection of cultured fish. The transfer of pathogens to wild fish can potentially affect the susceptibility of wild fish to disease and, in extreme cases, could affect wild fish abundance and local biodiversity.

The discharge of pathogens does not necessarily lead to disease. Infection with a pathogen is a much more common occurrence than disease. Infectious diseases result from the interaction of the health and immunological status of the host, the dose and virulence of the pathogen, and environmental conditions that affect host-pathogen interaction. The health of wild fish populations is influenced by natural factors, such as genetics, nutrition, environmental conditions, and the biology of pathogen and host organisms. Fish health is also affected by anthropogenic factors that put pressure on wild fish populations, including climate change, floods, drought, impoundments, dams, chemical contaminants, and fishing pressure.

The risk of disease caused by a particular pathogen is affected by the characteristics, distribution, survival, and fate of pathogens in the environment; factors affecting the route of pathogen transmission; the characteristics of pathogen exposure; the effect of environmental stressors; and the susceptibility of wild hosts to infection that can lead to disease. In general, the relative importance of each factor is difficult to understand and assess in wild populations, leading to a high degree of uncertainty associated with the risk of transfer of pathogens between cultured and wild fish populations.

Wild fish can be affected by the introduction of new pathogens attendant with the introduction of native or non-native species to an area for culture. These pathogens can infect wild individuals of the same or different species. Pathogens can also be introduced to wild fish through transportation of native or non-native species that are not intended for aquaculture.

Interactions of Escaped Fish with Wild Populations and Natural Ecosystems

Aquaculture production facilities are embedded in natural ecosystems with varying degrees of intimacy, depending on characteristics of the culture system. The hydrological connection to natural ecosystems can be remote and intermittent (*e.g.*, recirculating aquaculture systems) or intimate and continuous (*e.g.*, shellfish and net-pen systems). The risks of containment failure also vary with culture system type. As such, the probability of occurrence of escape is variable. Escapes can be defined as the unintentional and unplanned release of cultured animals or their gametes or offspring to the environment. Escapes do not include intentional releases of cultured animals for stock enhancement or ranching.

Mechanisms that lead to the escape of cultured organisms can be grouped into four classes. In each case, the mechanism leads to a breach in the barrier between a culture unit and the environment, allowing escapes to occur. The causes of containment failures that can lead to escape include 1) catastrophic natural events, 2) human error associated with facility operation, 3) vandalism or poaching, and 4) attacks by nuisance or predatory wildlife. Culturists have no control over natural disasters, but facilities can be constructed to withstand or minimize the risk of escapes by other mechanisms.

Escape of cultured animals is not desirable from an economic standpoint because escapes represent a loss of potential revenue. Therefore, there is a strong incentive to construct and manage production units to minimize losses from escapes, disease, poaching, and other causes. Above and beyond the undesirable economic impact on producers, escaped culture organisms can alter the physical environment and the structure of ecological communities, cause genetic changes in stocks of wild conspecifics, introduce parasites and diseases, and cause various socioeconomic impacts.

Escaped individuals may interbreed with natural populations, although interbreeding requires the satisfaction of several criteria. First, cultured organisms must be sexually mature or, if not mature, they must survive to maturity. The timing of the escape event must also coincide with the natural breeding season of the wild population; in other words, sexually mature individuals in the wild population must be receptive to escaped fish.

Intraspecific hybridization between escaped and wild conspecifics can reduce the genetic variance, and presumably fitness, of wild populations through outbreeding depression. This may increase the vulnerability of natural populations to environmental change from the loss of genetic differences between cultured and wild (natural) populations. The effect of inbreeding will depend on the degree of local adaptation. Geographically structured populations are more susceptible to the effects of interbreeding than undifferentiated stocks.

The principles of interaction between wild and escaped fish have been reviewed by Youngson *et al.* (2001). The overall concern is related to a reduction in fitness and a reduction in the effective population size of wild populations.

First, the genetic profile (*i.e.*, allele frequencies) of wild populations can change from the transfer of genes from cultured stocks into wild populations. Second, the genetic profile of cultured stocks may contain alleles that are rare or unusual in wild populations but were selected on the basis of performance traits, and that are transferred to wild stocks. Third, the abundance of wild populations may decrease if escaped fish displace wild fish, but then demonstrate low reproductive fitness. Although all of these mechanisms of interaction are plausible, demonstrating a measurable effect on the fitness of wild populations is difficult.

The consensus view is that the risk of adverse genetic effects on natural populations of conspecifics is not as severe as potential ecological effects of escapes. The intensity of the ecological impacts of escapes is a function of the number of organisms that escape, the timing of an escape event, the biology and life history of the escaping species, interactions of escaped organisms with wild populations, and ecological characteristics of the host ecosystem. Even non-reproductive animals can have ecological impacts if they escape in sufficient numbers.

Potential adverse impacts of escapes on natural populations may include any or all of the following:

- Competition with natural populations for habitat, food resources, nesting or spawning sites, or mates;
- Predation on endemic fish populations if escaped fish are piscivores;
- Predation on endemic fish populations if escaped fish attract predators;
- Transmission of pathogens or parasites with deleterious effects on wild populations;
- Amplification of the effect of endemic pathogens;
- Alteration of habitat, such as changes in turbidity and substrate; and
- Colonization of habitat, with negative effects on wild populations, and in the worst case, displacement of native species.

Overall, the potential outcomes of escapes may include a reduction in fitness in wild con-specific populations and a reduction in fitness in other wild populations. Many of the adverse impacts described above are exacerbated when invasive species escape from aquaculture facilities. Although only one of many pathways for the introduction of potentially invasive species, aquaculture has played an important role in the introduction of non-native species, especially in freshwater. The majority of introductions of non-native species since the 1970s can be attributed to aquaculture and arguably the introduction and establishment of invasive species from aquacul-ture has had the greatest adverse effect on the environment of all potential impacts. In any case, the effects of escapes by cultured species on biodiversity are difficult to partition from the effects of habitat degradation or loss and other causes of biodiversity reduction.

Collection of Wild Aquatic Animals for Spawning or Stocking

There is a long history in aquaculture of capturing wild larvae and juveniles for stocking into culture units for growth to market size. The culture of milkfish, mullet, penaeid shrimp, and many other species was long dependent on wild stocking material. Capturing wild juveniles is usually an interim step in the development of a species for culture. Once the techniques for spawning and early larval rearing are developed to close the life cycle, juveniles can be reliably produced in hatcheries, allowing the development of a sustainable aquaculture industry for that species. Most species in freshwater aquaculture are now spawned in captivity. The dependence on wild-caught stocking materials is becoming much less common, although reliance is still strong in brackishwater and marine aquacul-ture. The capture-based aquaculture for eels, groupers, tunas, and yellowtails remains dependent on capture of wild juveniles.

There are several problems related to the use of wild juveniles for stocking. Collection of juveniles may place additional pressures on wild stocks of the target species, reducing recruitment to natural fisheries if density-dependent mechanisms regulate abundance. Furthermore, depending on the homogeneity of aggregations of juveniles, individuals of non-target species of commercial or ecological importance may be collected incidentally as by-catch, placing pressure on wild stocks of those species. For example, in the Philippines, milkfish fry constitute only 15 per cent of collections and the 85 per cent of by-catch is discarded.

The effect of capture of wild fish as juveniles on target or non-target wild populations will depend on the size- and age-structure of the population, population size, age at maturity, natural and fishing mortality rates, and quality and availability of suitable habitat. Selective collection of wild juveniles can also have a negative effect on fish community structure, depending on the trophic role of the species and the complexity of the community structure. Depending on the gear employed, juvenile collection can result in physical disturbance of the substrate, impacts on benthic communities, and in the worst case, destruction of habitat. Finally, wild seed are undesirable from a biosecurity standpoint because the health status of wild-caught juveniles is questionable. Shrimp producers now perceive that postlarvae produced in a hatchery are less susceptible to important diseases than wild postlarvae.

In the culture of some species, sexually mature broodstock are collected from the wild and induced to spawn in hatcheries. Broodstock collection can also place pressure on wild fish resources. For example, in Thailand, Vietnam, and some other Asian countries, the availability of wild-caught black tiger shrimp (*Penaeus monodon*) for broodstock has declined drastically in recent years. Adult *P. monodon* broodstock are now rare in trawl catches. Capture of mature adults for use in shrimp hatcheries likely has been one of the reasons for the decline. This decline in *P. monodon* broodstock has been a major reason why shrimp producers in Thailand and other Asian nations have imported non-native, farm-reared broodstock of the Pacific white shrimp (*Litopenaeus vannamei*). This species has replaced much of the *P. monodon* culture in Thailand.

Predator Control and Other Effects on Wildlife

The concentration of potential food resources in aquaculture facilities provides attractive foraging opportunities for birds and other predators of cultured animals, which are viewed as a nuisance by aquaculture producers. The construction of aquaculture ponds has created new aquatic habitat while suitable habitat for wildlife with an aquatic orientation, especially migratory waterfowl, has decreased elsewhere. Ponds are attractive to waterfowl and migratory birds as sources of food and shelter. Aquaculture structures such as net pens and shellfish rafts create a complex, three-dimensional environment in the water column that attracts and aggregates wild fish, their predators, birds, turtles, and marine mammals.

Aquaculture producers use a range of non-lethal and lethal techniques to control preda-tion of cultured animals. Most methods are intended to harass wildlife and induce them to move elsewhere. Although lethal controls are used occasionally, it is unlikely that lethal control will have negative effects on wildlife populations, especially of those for which aquaculture has allowed an expansion of population abundance. Most wildlife is protected by international conventions and national regulations. It is possible that location in migratory pathways and near breeding grounds could cause disruption to wildlife populations, but it is unlikely that these sites would be selected or permitted for aquaculture.

There is some risk that wildlife may become entangled or trapped in nets, ropes, and floating or submerged structures for fish and shellfish. This could lead to drowning or starvation, but properly designed, constructed, and managed facilities do not present a major risk of entrapment or entanglement. In sum, there are few negative interactions with wildlife and the probability of negative consequences is very low.

GLOBAL ENVIRONMENTAL THREATS TO FISHERIES AND AQUACULTURE

We live on a human-dominated planet, and the momentum of human population growth, together with the imperative for further economic development in most of the world, ensures that our dominance will increase.

SUSTAINABLE DEVELOPMENT AND MANAGEMENT OF AQUACULTURE AND FISHERIES SYSTEMS

The sustainable development and management of aquaculture and fisheries systems can only occur if these activities are well planned and integrated into the natural and social resource, ecosystems, and farming systems contexts of the larger global context of which they are a part. Population and natural resource constraints in a crowded future demand that aquaculture "fits" as a part of a larger strategy for the non-consumptive, multiple uses of water; and that fisheries be managed sustainably as part of the larger trends affecting the marine environment. In the past 50 years there has been a massive migration—called the "greatest human migration of all time"—from rural, inland areas to the world's coasts, resulting in about 60 per cent of the Earth's people living within 100 km of the coast. Massive growth of coastal cities has raised issues of the future survival of coastal and estuarine ecosystems and habitats, and has put at risk the livelihoods of millions of people in traditional coastal communities that depend upon the sustainable capture and culture of aquatic living resources. Upwards of 35 per cent of the primary production of the temperate continental shelves is being harvested in fisheries, or discarded back to the sea as waste ("bycatch"). Coastal margins and oceans are among the most heavily used and modified areas of the planet, suffering disproportionate amounts of habitat destruction and pollution.

Humans have removed about 50 per cent of the world's mangroves. Human activities in the coastal zone deliver sewage, solid wastes, refuse (marine debris), sediments, dust, pesticides, and oil hydrocarbons to rivers, estuaries, and coastal areas. Assessing the range, magnitude and delivery of land-based sources of pollution to coastal oceans is a major global effort. It is estimated that about 80 per cent of marine pollution originates from land-based sources and activities.

Inputs of nutrients to the coastal zone from development and agriculture have caused an increase in toxic algae blooms, some of which have caused human disease and neurological impairment. In Shanghai, China, an estimated 13,000 factories discharged 10.55 million tons of waste in 1989, nearly twice the amount in 1980; about one-fifth was toxic waste. The Songhua River in China contains 10 tons of mercury and has waterborne mercury concentrations higher than those reported after the disaster in Minamata Bay, Japan. About one-third of China's coastal waters are polluted with oil; mercury and cadmium have been detected in 60 per cent and 33 per cent of seawater samples, respectively. In Poland, 92 per cent of the nation's rivers are "beyond classification", meaning that the level of pollution is greater than that described by any existing pollution category.

Humanity is the now one of the major driving forces in the Earth's hydrological cycle, using more than half of the world's freshwater run-off. Agricultural water use accounts for about 75 per cent of total global consumption, mainly through crop irrigation, while industrial use accounts for about 20 per cent, and the remaining 5 per cent is used for domestic purposes. Most of the world's rivers are dammed—there are 36,000 dams—and the number of dams is increasing.

Many major rivers (Colorado, Nile and Ganges) are so heavily used that little or no water reaches their deltas and the sea. Only 2 per cent of rivers in the USA run free. Major inland water areas (Aral Sea, Lake Chad, etc.) have nearly dried up or been greatly reduced to small lakes. It is estimated that two out of every three people will live in water-stressed areas by the year 2025. In Africa, it is estimated that 25 countries will be experiencing water stress (below 1,700 m^3 per capita per year) by 2025. About 450 million people in 29 countries suffer from water shortages. Clean water supplies and sanitation are major problems in many parts of the world, with 20 per cent of the global population lacking access to safe drinking water. Water-borne diseases from fecal pollution of surface waters are a major cause of illness in developing countries. Polluted water is estimated to affect the health of 1.2 billion people, and contributes to the death of 15 million children annually.

Losses of biodiversity are occurring at alarming rates. Rates of loss are estimated at 100—1000 times greater than natural rates of extinctions. Extinction rates are even higher on remote islands, such as Hawai'i, USA. Due to the ease and rapidity of global transportation, the world's ecosystems are becoming "homogenized" due to "biological invasions": the exotic introductions of plants and animals, bacteria and other pests to natal ecosystems. After land transformations, exotic species impacts are the second major cause of species extinctions and population losses. About 11 per cent of birds, 18 per cent of mammals, 5 per cent of fish and 8 per cent of the Earth's plants are threatened with extinction, and exotic species have contributed greatly to their demise.

Given these immense global challenges, Vitousek *et al.* recommend:

- A slowing the rate of population growth since humanity's growth drives all resource use and waste generation,
- A reduction in the rate of human impacts since ecosystems can react to lower rates of impacts and stabilize with moderate levels of human-induced changes,
- Accelerating ecosystem level research and management understanding. Human impacts must be included in all analyses of global and ecosystem change. Ecological science must include development of new methodologies to assist humanity in this global crisis, requiring development of true interdisciplinary environmental scholarship, and education of the "informed generalist."

THE GLOBAL IMPORTANCE OF FISHERIES AND AQUACULTURE

Fisheries play an important role in the world food economy. Fisheries are a source of employment for about 200 million people who depend directly upon ocean fishing for their livelihoods. Fish is the primary source of protein for some 950 million people worldwide and represents an important part of the diet of many more.

In less than 50 years, the world's average per capita consumption of fish has almost doubled. Globally, fish provide about 16 per cent of the animal protein consumed by humans, and are a valuable source of minerals and essential fatty acids.

Fish is the primary source of omega-3 fatty acids in the human diet. Omega-3 fatty acids are critical nutrients for normal brain and eye development of infants, and have preventative roles in a number of human illnesses, such as cardiovascular disease, lupus, depression and other mental illnesses.

The reported production of fish for direct human consumption doubled between 1950 and 1970, and has stabilized since then at an average of 9.0 to 10 kg of fish per capita, notwithstanding world population growth. Fish consumption per person is expected to continue to rise. Supply will probably be limited by environmental factors, and a likely range for demand is 150 to 160 million tons, or between 19 and 20 kg per person in 2030. Global increases in consumption of food fish will take place predominantly in the developing countries, where population is growing and higher incomes are allowing purchase of high value fisheries items for the first time by many people. However, fish production in least developed countries where fish protein is needed to prevent malnutrition is a key element of food security in these regions and a critical area where innovative programmes are needed to increase production.

Today, fishing is the largest extractive use of wildlife in the world. The value of world total fishery production in 1999 was US$ 125 billion. World production of fish, crustaceans, mollusks and plants reached 142 million tons in 2001. Capture fisheries production, which accounted for 66 per cent of the total, was 93.7 million tons, of which inland capture was 8.7 million tons, while aquaculture production was 48.4 million tons including plants. Marine and freshwater fish are also an increasingly important recreational resource, both for active users such as anglers and for passive users such as tourists, sports divers and nature-lovers.

PREDOMINATES IN CAPTURE FISHERIES AND AQUACULTURE PRODUCTION

Asia predominates in capture fisheries and aquaculture production, with China the leading nation worldwide. In 1999, over three-quarters (or 97 million tons) of the global production of fish, crustaceans and mollusks were utilized for direct human consumption. Fish utilized as raw material for the production of animal feed, *e.g.*, fish meal and fish oil, represented about one-quarter of the total fishery production in 1999; and this amount has remained relatively static for the past 15 years. Most growth in the fisheries sectors is projected to occur in developing countries, which will account for 79 per cent of food fish production in 2020.

Increased production of fish from aquaculture has occurred primarily as a result of increasing feed inputs into ponds and other production systems, thereby increasing yields per hectare by an order of magnitude compared to extensive production systems in which rearing water is fertilized only. Higher inputs mean two things to the aquaculture feed industry; more feed and higher quality feed. Currently, global feed production for farmed fish and crustaceans is approximately 13 million tons, and predictions are for feed production to increase to over 37 million tons by the end of the decade, an increase of 24 million tons. Feeds for salmonids and marine fish have always been complete feeds, *i.e.*, ones that supply all of the nutritional needs of the fish. Pond-reared fish, in contrast, obtain a significant proportion of their nutritional needs from pond biota. The degree to which feeds must supply essential nutrients to pond-reared fish increases as rearing densities increase beyond the capacity of natural foods in ponds to supply them. Fish farmers around the world have found that as they increase feed inputs, the biomass and economic yields from ponds increase as well. Thus, great areas of low-input, pond-based aquaculture mainly in Southeast Asia and China are being converted from low-input systems to high-input systems that depend upon high quality feeds to supply an increasing proportion of nutrients used by the fish.

The effects of the aquaculture industry's growth and of changes in feed input in pond-based aquaculture have been dramatic with respect to the use of marine proteins by the aquaculture feed industry. In the mid-1980s, less than 10 per cent of annual fish meal production was used by the aquaculture feed sector. Today, that proportion is over 40 per cent. Similarly, aquaculture now uses nearly 75 per cent of annual global fish oil production, up from less than 10 per cent seventeen years ago, but this change is mainly due to the adoption of high lipid feeds by the salmon farming industry, rather than increasing the inputs into pond-based aquaculture systems.

Fish meal and oil are produced from species of fish that are not generally utilized directly for human food, *e.g.*, herring and capelin in Norway and Iceland, sand eel in Denmark, capelin in South Africa, anchovies in Peru and northern Chile, jack mackerel in central Chile, sardines in Japan, and menhaden in the USA. The rapid increase in fish meal and oil use in aquaculture feeds over the past 15 years has not resulted in over-exploitation of the stocks of fish harvested primarily to produce fish meal and oil. Production of fish meal has averaged between 6-7 million tons per year since the mid-1980s, except in El Nino years, when production has been lower. Fish oil production has averaged between 1.2-1.3 million tons. No significant changes in annual harvest or fish meal/oil production associated with increased aquaculture production or increasing intensification of aquaculture and concomitant higher feed inputs is evident. Increased fish meal/oil use in the aquaculture sector has come at the expense of other uses.

Estimates of future protein requirements for aquaculture feeds depend upon future production from various segments of the aquaculture industry and on annual production levels of fish meal. As aquaculture production expands, it will be essential to replace portions of fish meal and oil in fish feed formulations with alternative ingredients derived from alternative sources, primarily grains and oilseed products. Because 70 per cent of the fish meal used in diets for fish and crustaceans is used to produce diets for salmonids, marine fish and shrimp, these production sectors are the focus of most research with respect to the use of alternative protein sources. Numerous studies have been conducted to evaluate the effects of replacing various percentages of fish meal in diets for these fish, and, without exception, none has successfully replaced 100 per cent of fish meal without reducing fish performance. At best, 50 per cent of fish meal in diets for salmon and trout can be replaced by soy protein concentrate, and 25-30 per cent with soybean meal. Similar findings have been reported in studies of wheat gluten meal, corn gluten meal, and rapeseed protein concentrate. For the most part, formulated fish diets are used to produce high-value fish for export. In underdeveloped countries, increased production of lower-value fish that are consumed by local populations will depend upon the use locally-available, low-cost feed ingredients that can be combined to produce prepared feeds to increase productivity of community ponds and waterways.

This is an area that requires substantial investigation and development, and is a logical target for USAID support. Another logical target for USAID efforts is the development and testing of feeds that are based upon sustainable plant-derived protein sources for use in developing countries where fish are grown both for domestic consumption and export, such as China. Partnerships with US commodity groups could be a mechanism to extend USAID efforts. It is critical that USAID become a leader in this area, given the perception that increased aquaculture production leads to higher fish meal use and greater pressure on stocks of fish that are captured to produce fish meals and oils.

Fisheries products have become the most internationally traded food, as some 37 per cent (by quantity) of all fish for human consumption is traded across borders. In 1999, international trade (in live weight equivalent) represented 34 per cent of the total production. In 1999, foreign trade earnings amounted to US$ 52.9 billion. Most fishery exports were destined to developed countries, with Japan the world's leading importer, and the USA second. The US balance of trade deficit in seafood products is approximately $9 billion.

Exports of fisheries products to the developed countries have become so lucrative that nations like Argentina, a traditional, globally important exporter of meat products and livestock have turned to seafood exports as the major source of foreign exchange earnings. In 2002, exports of raw and processed fish and shellfish from Argentina surpassed meat products and livestock, with beef earning US$ 574 million in export revenue, while seafood—mostly prepared products—earned US$ 714 million. Argentina's main consumers of domestic seafood products were Spain, Brazil and the US.

7

Biotechnology Applications in Aquaculture and Fisheries

Biotechnology provides powerful tools for the sustainable development of aquaculture, fisheries, as well as the food industry. Increased public demand for seafood and decreasing natural marine habitats have encouraged scientists to study ways that biotechnology can increase the production of marine food products, and making aquaculture as a growing field of animal research. Biotechnology allows scientists to identify and combine traits in fish and shellfish to increase productivity and improve quality. Scientists are investigating genes that will increase production of natural fish growth factors as well as the natural defence compounds marine organisms use to fight microbial infections. Modern biotechnology is already making important contributions and poses significant challenges to aquaculture and fisheries development. It perceives that modern biotechnologies should be used as adjuncts to and not as substitutes for conventional technologies in solving problems, and that their application should be need-driven rather than technology-driven.

The use of modern biotechnology to enhance production of aquatic species holds great potential not only to meet demand but also to improve aquaculture. Genetic modification and biotechnology also holds tremendous potential to improve the quality and quantity of fish reared in aquaculture. There is a growing demand for aquaculture; biotechnology can help to meet this demand. As with all biotech-enhanced foods, aquaculture will be strictly regulated before approved for market.

Biotech aquaculture also offers environmental benefits. When appropriately integrated with other technologies for the production of food, agricultural products and services, biotechnology can be of significant assistance in meeting the needs of an expanding and increasingly urbanised population in the next millennium. Successful development and application of biotechnology are possible only when a broad research and knowledge base in the biology, variation, breeding, agronomy, physiology, pathology, biochemistry and genetics of the manipulated organism exists. Benefits offered by the new technologies cannot be fulfilled without a continued commitment to basic research. Biotechnological programmes must be fully integrated into a research background and cannot be taken out of context if they are to succeed.

Indian fisheries and aquaculture is an important sector of food production, providing nutritional security to the food basket, contributing to the agricultural exports and engaging about fourteen million people in different activities. With diverse resources ranging from deep seas to lakes in the mountains and more than 10 per cent of the global biodiversity in terms of fish and shellfish species, the country has shown continuous and sustained increments in fish production since independence. Constituting about 4.4 per cent of the global fish production, the sector contributes to 1.1 per cent of the GDP and 4.7 per cent of the agricultural GDP. The total fish production of 6.57 million metric tonnes presently has nearly 55 per cent contribution from the inland sector and nearly the same from culture fisheries. Fish and fish products have presently emerged as the largest group in agricultural exports of India. The potential area of biotechnology in aquaculture include the use of synthetic hormones in induced breeding, transgenic fish,gene banking, uniparental and polyploidy population and health management.

BIOTECHNOLOGY IN FISH BREEDING

Gonadotropin releasing hormone (GnRH) is now the best available biotechnological tool for the induced breeding of fish. GnRH is the key regulator and central initiator of reproductive cascade in all vertebrates.It is a decapeptide and was first isolated from pig and ship hypothalami with the ability to induce pituitary release of luteinising hormone (LH) and follicle stimulating hormone (FSH). Since then only one form of GnRH has been identified in most placental mammals including human beings as the sole neuropeptide causing the release of LH and FSH. However,in non- mammalian species (except guinea pig) twelve GnRH variants have now been structurally elucidated, among them seven or eight different forms have been isolated from fish species. Depending on the structural variant and their biological activities, number of chemical analogues have seen prepared and one of them is salmon GnRH analogue profusely used now in fish breeding and marked commercially under the name of 'Ovaprim' throughout the world. The induced breeding of fish is now successfully achieved by development of GnRH technology.

TRANSGENESIS

Transgenesis or transgenics may be defined as the introduction of exogenous gene/ DNA into host genome resulting in its stable maintenance, transmission and expression. The technology offers an excellent opportunity for modifying or improving the genetic traits of commercially important fishers, mollusks and crustaceans for aquaculture. The idea of producting transgenic animals became popular when first produced transgenic mouse by introducing metallothionein human growth hormone fusion gene (mT-hGH) into mouse egg, resulting in dramatic increase in growth. This triggered a series of attemptson gene transfer in economically important animals including fish.

The first transgenic fish was produced in China, who claimed the transient expression n putative transgenics, although they gave no molecular evidence for the integration of the transgene.

The technique has now seen successfully applied to a number of fish species. Dramatic growth enhancement has been shown using this technique especially in salmonids. Some studies have revealed enhancement of growth in adult salmon to an average of 3 P5 times the size of non- P transgenic controls, with some individuals, especially during the first few months of growth, reaching as much as 10 P30 times the size of the controls.The introduction of transgenic technique has simultaneously put more emphasis on the need for production of sterile progeny in order to minimize the risk of transgenic stocks mixing in the wild populations.

The technical development has expanded the possibilities for producing either sterile fish or those whose reproductive activity can be specifically turned on or off using inducible promoters. This would clearly be of considerable value allowing both optimal growth and controlled reproduction of the transgenic stocks while ensuring that any escaped fish would be unable to breed. An increased resistance of fish to cold temperatures has been another subject of research in fish transgenics for the past several years. Coldwater temperatures pose a considerable stressor to many fish and few are able to survive water temperatures much below 0-1°C. this is often a major problem in aquaculture in cold climates. Interestingly, some marine teleosts have high levels (10 P25 mg/ml) of serum antifreeze proteins (AFP) or glycoproteins (AFGP) which effectively reduce the freesing temperature by preventing ice-crystal growth.

The isolation, characterisation and regulation of these antifreeze proteins particularly of the inter flounder Pleuronectas americanus has been the subject of research for a considerable period in Canada. Consequently, the gene encoding the liver AFP from winter flounder was successfully introduced into the genome of Atlantic salmon where it became integrated into the germ line and then passed onto the off P spring F3 where it was expressed specifically in the liver.The introduction of AFPs to gold fish also increased their cold tolerance, to temperatures at which all the control fish died (12 h at 0° C; Wang *et al.*, 1995). Similarly, injection or oral administration of AFP to juvenile milkfish or tilapia led to an increase in resistance to a 26 to 13° C. drop in temperature. The development of stocks harbouring this gene would be a major benefit in commercial aquaculture in counties where winter temperatures often border the physiological limits of these species.

The most promising tool for the future of transgenic fish production is undoubtedly in the development of the embryonic stem cell (ESC) technology. There cells are undifferentiated and remain totipotent so they can be manipulated *in vitro* and subsequently reintroduce into early embryos where they can contribute to the germ line of the host. This would facilitate the genes to be stably introduced or deleted.Although significant progress has been made in several laboratories around the world, there are numerous problems to be resolved before the successful commercialisation of the transgenic brood stock for aquaculture. To realise the full potential of the transgenic fish technology in aquaculture, several important scientific break P through are required.

There include:

- More efficient technologies for mass gene transfer
- Targeted gene transfer technologies such as embryonic stem cell gene transfer
- Suitable promoters to direct the expression of transgenes at optimal levels during the desired developmental stages
- Identified genes of desireable traits for aquaculture and other applications
- Informations on the physiological, nutritional, immunological and environmental factors that maximize the performance of the transgenics of the transgenics and
- Safety and environmental impacts of transgenic fish.

CHROMOSOME ENGINEERING

Chromosome sex manipulation techniques to induce polyploidy (triploidy and tetraploidy) and uniparental chromosome inheritance (gynogenesis and androgenesis) have been applied extensively in cultured fish species. These techniques are important in the improvement of fish breeding as they provide a rapid approach for gonadal sterilisation, sex control improvement of hybrid viability and clonation. Most vertebrates are diploid meaning that they possess two complete chromosome sets in their somatic cells. Polyploidy individuals possess on or more additional chromosome sets, bringing the total to three in triploids, four in tetraploids and so on.

Induced triploidy is widely accepted as the most effective method for producing sterile fish for aquaculture and fisheries management. The methods used to induce triploids and other types of chromosome set manipulations in fishes and the applications of these biotechnologies to aquaculture and fisheries management are well described. Tetraploid breeding lines are of potential benefit to aquaculture, by providing a convenient way to produce large numbers of sterile triploid fish through simple interploidy crosses between tetraploids and diploids. Although tetraploidy has been induced in many finfish species, the viability of tetraploids was low in most instances.

In teleosts, technique for inducing sterility include exogenous hormone treatment and triploidy induction. The use of hormone treatements, however could be limited by governmental regulation and a lack of consumer acceptance of hormone treated fish products. Triploidy can be induced by exposing eggs to physical or chemical treatment shortly after fertilization to inhibit extrusion of the second polar body triploid fish are expected to be sterile because of the failure of homologous chromosomes to synapse correctly during the first meiotic division. Methods of triploidy induction in clued exposing fertilized eggs to temperature shock (hot or cold), hydrostatic pressure shock or chemicals such an colchicines, cytochalasin-B or nitrous oxide.

Triploid can also be produced by crossing teraploids and diploids. Tetraploid induction involves fertilizing eggs with normal sperm and exposing the diploid sygote for physical or chemical treatment to suppress the first mitotic division. Gynogenesis is the process of animal development with exclusive maternal inheritance. The production of gynogenetic individuals is of particular interest to fish breeders because a high level of inbreeding can be induced in single generation.

Gynogenesis may also be used to produce all P female populations in species with female homogamety and to reveal the sex determination mechanisms in fish. It is convenient to use all female gynogenetic progenies (instead of normal bisexual progenies) for sex inversion experiments. Methodologies combining use of induced gynogenesis with hormonal sex inversion have been developed for several aquaculture species.

Androgenesis is the process by which would have commercial application in aquaculture. It can also be used in generating homozygous lines of fish and in the recovery of lost genotypes from the crypreserved sperms. Androgenetic individuals have been produced in a few species of cyprinids, cichlids and salmonids.

BIOTECHNOLOGY AND FISH HEALTH MANAGEMENT

Disease problem are a major constraint for development of aquaculture. Biotechnological tools such as molecular diagnostic methods, use of vaccines and immunostimulants are gaining popularity for improving the disease resistance in fish and shelfish species world over for viral diseases, avoidance of the pathogen is very important. In this context there is a need to rapid method for detection of the pathogen.

Biotechnological tools such as gene probes and polymerase chain reaction (PCR) are showing great potential in this area. Gene probes and PCR based diagnostic methods have developed for a number of pathogens affecting fish and shrimp. In case of finfish aquaculture, number of vaccine against bacteria and viruses have been developed. Some of these have been conventional vaccines consisting of killed microorgansism but new generation of vaccine consisting of protein subunit vaccine genetically engineered organism and DNA vaccine are currently under development.

In the vertebrate system, immunisation against disease is a common strategy. However the immune system of shrimp is rather poorly developed, biotechnological tools are helpful for development of molecule, which can stimulate this immune system of shrimp. Recent studies have shown that the non- specific defence system can be stimulated using, microbial product such as lipopolysacharides, peptidoglycans or glucans. Among the immunostimulants known to be effective in fish glucan and levamisole enhance phagocytic activities and specific antibody responses.

CRYOPRESERVATION OF GAMETES OR GENE BANKING

Cryopreservation is a technique, which involve long-term preservation and storage of biological material at a very low temperature usually at -196 C, the temperature of liquid nitrogen. It is based on the principle that very low temperature tranquilise or immobilise the physiological and biochemical activities of cell, thereby making it possible to keep them viable for very long period.

The technology of cryopreservation of fish spermatozoa (milt) has been adopted for animal husbandary. The first success in preserving fish sperm at low temperature was reported by Blaxter who fertilizes Herring eggs with frozen thawed semen.The spermatozoa of almost all cultivable fish species has now been cryopreserved. Cryopreservation overcomes problems of male maturing before female, allow selective breeding and stock improvement and enables the conservation.

One of the emerging requirement for that can be used by breeders for evolving new strains. Most of the plant varieties that has been produced are based on the gene bank collections. Aquatic gene bank however suffers from the fact that at present it is possible to cryopreserve only the male gametes of finfishes and there in no viable technique for finfish eggs and embryos. However, the recent report on the freezing of shrimp embryos. However, the recent report on the freesing of shrimps embryos by subramoniam and newton and Diwan and kandaswami look promising. Therefore, it is essential that gene banking of cultivated and cultivable aquatic species be undertaken expeditiously.

FISHERIES AND AQUACULTURE TECHNOLOGY

The technology employed in aquaculture has developed over many centuries but has done so more rapidly during the last half century. Aquaculture systems, and the technology used, vary from very simple systems, used for family ponds in tropical countries, where production is for domestic consumption, to high technology systems, such as intensive closed systems, like those used for rearing stripped bass. Herbivorous and filter feeding fish, reared mostly in simple systems of small freshwater ponds, however, account for about half of global aquaculture production.

Much of the technology used in aquaculture is relatively simple, amounting to small modifications that improve the growth and survival rates of the target species, such as providing additional food, adding seed animals collected elsewhere, managing water exchange to maintain adequate oxygen levels, and protecting the stock from predators. Greater understanding of complex interactions between nutrients, bacteria and cultured organisms, together with advances in hydrodynamics applied to pond and tank design, have enabled the development of closed systems. These have the advantage of isolating the aquaculture systems from natural aquatic systems, thus minimizing the risk of disease or genetic impacts on the external systems.

Developments in engineering, some learnt from offshore oilrig construction, increase the possibilities for offshore aquaculture using robust cages. Sea ranching, the release of young fish into the wild to improve the harvest in capture fisheries, has also made a start but its long term viability is still to be assessed. Major advances are also being made in the technology of the production of aquafeeds, which generally require the combining of a large number of ingredients into very small feed pellets.

AQUACULTURE ENGINEERING

Aquaculture draws on well-established engineering fields for most of the design and construction needs of its production facilities. Building earth ponds is similar to building roads - a knowledge of the characteristics of soils and the limits of safe design are the basis of good construction. Similarly, the buildings used for hatcheries and other support activities are no different from those common in the housing, agricultural and commercial sectors.

Sometimes ponds are lined with plastics or other impermeable materials, and here the techniques are similar to those for civil structures such as potable water reservoirs or sludge tanks. The design and installation of water control gates, including in unstable soils, benefits from the long experience in this field in the agriculture and irrigation sectors. An understanding of hydrodynamics allows ponds and tanks to be built with good water circulation, oxygen mixing and without 'dead spots' where sediments might accumulate and cause health problems to fish.

For installations in the sea, the situation is somewhat different and many of the important engineering solutions, such as for fish cages or suspended shellfish growout systems, have had to be developed by aquaculturists themselves. They have benefited however from the accumulated knowledge of seafarers in general and fishermen in particular, in the design and operation of mooring and buoyage systems. More recently, when fish farmers have turned their attention to how to operate fish cages in locations further offshore where seas are rougher, the experience of the oil exploration industry has proven very valuable.

Techniques have been developed in recent years for the production of fish and other aquatic products in closed recirculation systems. To make these work, aquaculturists have needed to develop a knowledge of the biological processes operating - such as how bacteria can be used to neutralise and re-cycle the nitrogenous waste products produced by growing fish - and how to engineer the systems to meet the biological requirements. Knowledge of bio-engineering from the waste treatment and water treatment industries has made contributions to the development of closed aquaculture systems and there is probably more that could be usefully transferred from the sewage treatment sector to help solve problems in fish rearing.

Modern materials have brought many benefits to aquaculture. For instance, custom made plastic joints have simplified the construction of sea cages, and made them more reliable in high stress conditions. Experiments with huge free-floating or sunken net cages operated in the open ocean were begun several decades ago, for instance in the Caspian Sea. These showed some promise, but more reliable construction materials will make the farming of fish in such structures increasingly feasible. Modern materials and production methods have been important also, in the construction of plastic filter substrates for indoor recirculating systems. The fine detail of these has been found to make substantial differences to the efficiency of biological filters.In shrimp farms, specially designed matting materials that stand upright on pond bottoms, with a structure that promotes the growth of the small animals and plants that the shrimp can thrive on, have recently been developed and shown to boost production.

Engineering skills are important in the design of most aquaculture facilities and good engineering can affect the efficiency and economics of production. If capital costs can be minimised while still maximising productivity and reducing risk, the farming operation will be more profitable.

Aquaculturists have proven very innovative over the past fifty years, constantly developing new technologies to support their farming operations. As new production methods and species for farming are developed, the engineering solutions needed to support them will continue to evolve.

AQUACULTURE FACILITIES

Aquaculture began with man making small modifications to natural habitats so as to improve the survival and growth of target species. Some of the oldest examples are in the rearing of freshwater fish in ponds, which has been practiced for thousands of years in Asia and at least for many centuries in Europe. The simple act of placing a mesh barrier across the outlet of a small pond or lake to prevent fish from escaping can make a big improvement in the supply of food.

Similarly, along coasts, some tidal lagoons can be easily turned into ponds. Closing off such naturally occurring water bodies was the start, centuries ago, of much fish and shrimp aquaculture in Asia and, in modern times, in South America. Removing predators and improving the conditions within the pond, (for instance by providing more area of preferred water depth), supplying additional food and, later, by adding seed animals collected outside, were further steps that moved aquaculture production close to where it is today. The farming of seaweeds and molluscs (oysters, clams, mussels etc) developed similarly, as people made improvements such as providing more settlement areas for the young, or the removal of predators from the growing area.

Farmers had a natural desire to improve the productivity of their systems and, as knowledge grew, they learned to stock more animals, increase feeding and manage the exchange of water to maintain the conditions, such as adequate oxygen levels, that the animals needed to survive. The range of facilities used for aquaculture subsequently broadened for a number of reasons. Farmers encountered difficulties with more intensive rearing because of the uncontrolled influence of pond soils, local water quality, weather. Some of these problems could be resolved by rearing in ponds built of concrete, or lined with plastic, by bringing the ponds indoors under cover, or treating the water before flowing it to the culture ponds.

Secondly, as the naturally occurring ponds and lagoons all became used, prospective farmers had to take a broader approach and develop the technology and engineering to be able to use less naturally-favored sites. Net cages floating in protected coastal or inland waters were developed for fish culture, or fish were stocked in fenced areas of the sea or of large lakes. At the same time, the need for supplies of young animals (fry, seed) to stock these systems led to the development of hatchery techniques and dedicated hatchery facilities. For most species this aspect of production has proved more successful when conditions can be more closely controlled, for instance in indoor concrete and fibreglass tank systems, rather than in outdoor ponds. Often, the younger stages of aquatic animals are more sensitive than adults to physical and chemical conditions and these have to be managed within a smaller range, if production is to be successful. Thus hatchery facilities developed as a separate branch of the industry.

More recently, our knowledge has improved greatly regarding the complex interactions that occur in a rearing system, between nutrients, bacteria and the cultured organism. This and technological developments have allowed many aquatic organisms to be reared in completely closed recirculating facilities, including the farming of marine organisms at locations far from the sea. Closed systems have the added advantage of offering greater protection from the danger of disease entering from the natural environment and also of minimizing adverse effects of the production system on that external environment.

Technology has also begun to open up the possibilities of growing fish in enclosures in the open ocean, something that could one day transform the nature of human food production on the planet. With 70 per cent of the earth's surface covered by water, the potential is clear. Earlier technology has restricted the cage farming of fish to sheltered coastal waters. Here the number of available sites is limited, environmental damage is more likely and conflicts exist with other users. As cages are developed that can withstand the demanding conditions of the open ocean, or which can be operated below the ocean surface, farming could move further offshore.

A small start has been made, notably in Japan, through so-called 'ranching' programmes', the farming of the sea without enclosures where fry are released in large numbers into the ocean with the aim of improving returns from the capture fishery. Sometimes artificial reef structures are created underwater as well, to increase the available habitat and natural food for the fish.

DEVELOPMENT OF AQUACULTURE SYSTEMS

Aquaculture - the growing of aquatic animals and plants - covers a wide range of species and methods. Some of the simplest production systems are the small family ponds in tropical countries where carp are reared for domestic consumption. At the other end of the scale are high technology systems, such as the intensive indoor closed units used in the USA for the rearing of striped bass or the sea cages used in Chile and Europe for growing salmon and bream. Nearly half of the world's aquaculture however is of herbivorous and filter-feeding fish such as carp. Practised in freshwater, in ponds made of earth, and often managed by a single household, production systems are simple and in many cases have changed little over many centuries. A farmer will excavate a small pond near his house; sometimes, as for instance in the flood river deltas of countries like Vietnam, using the spoil to build a raised foundation for the house itself. The pond will typically be fed by natural rainfall or groundwater, or sometimes by diverting a nearby stream or irrigation canal.

Some stocking occurs naturally, but more commonly a farmer will buy young fish from a breeder and stock them in his pond, often after holding them for a time in a small floating net known as a 'hapa' to check that they are in good condition before release in the pond. Production will improve if the farmer can stock a good balance of different species to make the best use of the varied kinds of food available in the pond. Typically the fish will be fed with household or agricultural by-products, and harvested when the family needs a meal. Stocking and harvesting is often continuous, where knowledge of the best stocking level is learned from experience and where the pond is not drained for many years. For most such farmers, fish production is a secondary activity, a useful additional source of protein to add to the supply or income from his main agricultural or commercial activities.

Outdoor freshwater fish farming is practised commercially in many countries, including in the developed nations. The density of stocking may be much higher, the control of feeding, water quality and fish health more closely monitored, but even in intensively operated systems, the key parameters that need to be monitored and balanced for success are similar to those in the traditional household pond system.

Over the last half-century, systems for the indoor rearing of many species of fish have also been developed. Holding of fish in controlled conditions indoors has been important in the development of seed production (hatcheries). Also, as knowledge has increased of nutrient cycles, bacterial action and water chemistry, it has become possible to rear fish, both for food and for ornament, in 'closed' indoor systems, where the water is recirculated. Passing the water through filters that use bacteria to naturally break down and recycle the waste products produced by the fish allows a production system to be run in almost total isolation from the outside. Such systems have been important in the control of disease and also in maintaining the stable conditions that some species need to flourish and reproduce.

A fifth of current world aquaculture production is of plants - mainly seaweeds, that for the most part are grown on the seabed or on raft or racks in shallow coastal waters and used directly for food or for the production of alginate or carageenan (agar-agar). This sector of aquaculture is another largely run by small-scale growers, mainly in Asia but in some parts of the South America, who attach seedlings of marine plants to simple structures built close to shore and constructed from natural materials and then harvest the plants once they have grown.

A further fifth of world production is of molluscs such as oysters, clams and mussels. Again the majority of production is small-scale and by coastal people who often combine fishing activities with farming. Oysters and mussels are mainly grown on structures built above the seabed - poles or racks on the shore, or ropes suspended from rafts or floating lines. The farmer's role is to supply suitable places for seed to settle or to add to natural production by bringing seed from a hatchery - and then to maintain the conditions of waterflow and freedom from predators that the shellfish need to grow. Most commercially grown molluscs feed on microscopic algae floating in the water, so the farmer does not need to provide any feed.

Fish are also grown in coastal areas, in ponds or in floating cages. As technology advances, there is the potential to develop systems for rearing fish in the open ocean, either in sturdy cages or by so called 'ranching', where young fish are released to the wild and then collected by normal fishing or by training them to respond to specially generated sounds.

The tropical coastal zones of Asia and Latin America are where most shrimp farms are found. Shrimp farming is mainly carried out in earth-bottomed ponds built on flat land close to the sea. Shrimp are raised in ponds at a range of different densities. At low densities the systems need only limited inputs of feed and fertilizers. As the density increases, more feed has to be supplied and at the higher end of the range, so-called 'intensive' farming, machines have to be put in the ponds to mix and aerate the water.

BIOTECHNOLOGICAL APPLICATION

Biotechnology - any technological application that uses biological systems, living organisms, or derivatives thereof, to make or modify products or processes for specific use 1994 UNEP Convention on Biological Diversity - has a wide range of useful applications in fisheries and aquaculture. It creates opportunities, for instance, to increase growth rate in farmed species, boost the nutritional value of aquafeeds, improve fish health, help restore and protect environments, extend the range of aquatic species and improve the management and conservation of wild stocks. Some biotechnologies are simple with a long history of application such as the fertilization of ponds to increase feed availability. Others are more advanced and take advantage of increasing knowledge of molecular biology and genetics, *e.g.* genetic engineering and DNA disease diagnosis.The field of genetic biotechnology similarly ranges from simple techniques such as hybridization, to more complex processes such as the transfer of specific genes between species to create GMOs (genetically modified organisms).

Over the years, our knowledge of fish breeding requirements has improved and the ability to induce artificial breeding developed through the use of natural or synthetic hormones and/or environmental manipulations. (For example changing photoperiod or water temperature can induce some fish to spawn). These have been key factors facilitating the application of more advanced biotechnologies.

Selective breeding, the maintenance of stocks genetically improved by chromosome manipulation, line crossing, and sex reversal all depend on the controlled breeding of farmed species. These improvements in reproductive technologies have also assisted aquaculturists greatly in their efforts to domesticate aquatic animals. In addition, by making it possible to remove the natural constraints and timing of breeding, farmers are able to mate many more species at times that are most beneficial, and thus helping to ensure a steady and consistent supply of fish for consumption.

AQUACULTURE EXTENSION

Aquaculture contributes significantly to the rural economy of most of the Asian and other developing countries by providing part- and full-time occupation to the farmers, fishermen and landless agricultural labourers. India and other developing countries of the South Asian region are endowed with ample water resources in the shape of freshwater ponds and tanks for fish culture, but these are not under effective and optimum utilization in spite of highly developed available technologies. Research results have shown excellent production potential as well as economic viability of such technologies, but until these technologies are successfully transferred to the beneficiaries, the desired objective cannot be achieved. Like agriculture, aquaculture is also an agroclimatic-based technology which, when developed in one agroclimatic region, may need modifications and refinement for adoption to another region. Agriculture extension thus involves not only the extension of aquaculture technology, but also certain levels of adoptive research in a particular field environment before it is launched for large-scale extension.

It is a two-way education process in which both scientists and farmers contribute, receive and interact with the involvement of extension workers as a link between the two and a catalyst as well (model). In other words, it is a non-formal adult education programme for educating and training the rural mass to acquire suitable fish farming skills and capabilities with a view to boosting fish production efficiency and the socio-economic condition.

Aquaculture extension is basically an educational process by means of which scientific and technological knowledge of aquaculture is carried to the farmers to upgrade their existing operation and farm management skills. The philosophy behind this process is to change the altitude, enhance the skill and knowledge of the fish farmers to upgrade their aquaculture practice. It also aims at binging maximum possible unutilized and under-utilized water areas under modern fish culture operation so as to raise the standard of living of the fish farmers through improving productivity and profitability. Apart from achieving its own target its overall objective is also to signifiantly contribute towards rural development by improving rural economy, creating additional gainful employment opportunities, fighting malnutrition and preventing rural exodus.

LAUNCHING AQUACULTURE TENSION PROGRAMME

Any aquaculture extension programme is designed based upon broad national consideration to achieve national goals and targets viz-a-viz local considerations to achieve short-term objectives such as application of composite fish culture in undrainable ponds to improve the aquaculture production level. A local aquaculture extension programme is relatively more definite in terms of scope and target. Like any other aquaculture extension programme, there should be three sequential steps for the dissemination of fish culture technology in undrainable ponds. They are as follows:

- Programme planning
- Programme implementation
- Programme evaluation

Programme Planning

While planning the dissemination of fish culture technology one should always bear in mind that the programme should be a self-regenerating production endeavour and once it is stimulated should continue on its own with a changed attitude and active participation of the recipient. This involves situation-specific strategies. The main components of programme planning are pre-adoption survey of the area, situation analysis, setting programme goals and finally designing strategies in a sequential manner.

Village Survey

Fish culture is basically a rural farming system and hence village survey is the most common method for identification of the difficulties faced by the farmers and to find out the scope and suitability of a specific technology needing to be transferred. The main objective is to get an overall picture of the village and the villagers, their attitude, values, together with their socio-economic conditions and also to locate and assess the available freshwater resources.

It also helps to identify the local institutions, village leaders, progressive farmers, school teachers, village level workers in order to design the most feasible extension strategy and also to establish a permanent rapport to strengthen the extension services.

At micro level it provides information about the socio-economic conditions of individual fish farmers, the pattern of fish farming and the technological gap.

The village selected for the survey should be such that it may represent the locality. Regular contact with important and progressive farmers of the village should be maintained. They should be informed about the objectives of the survey proposed to be undertaken.

Interviews with these persons will provide an overall picture of available natural and human resources and possible areas for development. Finally detailed relevant information may be collected from individual pond owners/ fish farmers and fishermen through personal interviews/questionnaires.

Resource Inventory

- *Availability of water resources:* Various types of water resources are available for fish culture but usually all of them are not fully utilized. Large, medium and smaller types of water bodies are generally available in villages which may be suitable for fish culture. Many small water bodies are found fully shaded by large marginal trees and thereby lying unproductive. Some unconventional types of water areas with potentiality for intensive aquaculture are also available.

  Canal/road and village side small and large ditches, pits emerging due to construction of mud houses etc., are some of the unnoticed

and untrapped potential aquaculture resources suitable for seed production and short-term fish rearing.
Low-lying and swampy areas which are formed naturally due to human activities are also potential sites for undrainable ponds for fish culture.

- *Availability of human resources:* It is a well-known fact that the majority of the people in developing countries live in villages and most of the rural population depend upon agriculture, aquaculture, livestock farming and other allied activities for their livelihood. Human resources are the vital inputs in rural aquaculture development. Rural areas have vast potential of unutilized or underutilized human resources for both men and women, which can be effectively utilized in operating aquaculture.
- *Identification of possible constraints:* A village survey also offers an excellent opportunity to identify various constraints in the background of which an appropriate strategy can be suitably designed.

Financial

Farmers usually do not have surplus funds big enough to be diverted towards reclamation and renovation of existing watersheds as well as construction of new ponds. Initial expenditure for fish culture over fish toxicant, fish seed and supplementary feed is itself a considerably big amount to be exclusively borne by farmers themselves without any credit support. As such, possible sources for mobilizing credit facilities may be identified.

Improper Water Area Distribution Pattern

Like land distribution pattern, major water areas are usually found in the possession of medium and big farmers who bother least about fish culture and concentrate themselves mostly on agriculture, while small and marginal farmers have minimum water holdings at their disposal with adequate manpower potential to be utilized. In some areas most of the water bodies are vested to village institutions, local administrative bodies, etc.

Lack of Technical Knowhow

Several seasonal and perennial ponds without any proper embankments are found lying fallow in a derelict condition due to ignorance and lack of technical knowhow. In some cases farmers fail to follow-up the prescribed package of practices strictly and land themselves in a state of financial turmoil and lose confidence in the viability of newly developed fish farming technologies.

Lack of Stocking Materials and Other Material Inputs

Fish farmers usually face the biggest problem of unavailability of quality fish seed for stocking their pond. Paucity of quality fish seed in the locality force the farmers to stock their ponds without any consideration to proper stocking size, density, species, ratio, etc. At times, they procure riverine fish seed which is usually mixed with the seeds of predatory and weed fishes. Other material such as fish toxicants are usually localised in its availability. All such problems are also vital for deciding area specific extension strategies.

Marketing Problem

It is a general practice that the fish is sold to middlemen at the pond site who invariably pay lower prices. Due to the perishable nature of the commodity and fear of exploitation by the fish wholesellers, farmers prefer to sell the crop at their pond/farm sites even at relatively lower rates. Information related to marketing practices will add to the scope of the extension programme so that farmers may be educated in marketing management to avoid such exploitation.

Lack of Transport and Efficient Communication System

In remote villages of India and many developing countries where fish culture technology needs to be extended, proper transport and communication facilities are lacking.

Social and Administrative Problems

Ponds remaining unutilized and lying in derelict conditions are common sights in rural areas in spite of a certain level of fish culture know how available with the farmers. In most cases such conditions exist due to family rivalry and non-cooperation among the members of the owners especially when the water areas are under multi-ownership. Poaching and deliberate poisoning of the ponds to destroy the crop are also serious social problems.

In some areas fish culture is supposed to be of a low-caste profession, thus many efficient upper-caste prople remain reluctant to come forward for this venture. Local administration such as Panchayats and Block Level Development Departments are also not always suitable geared enough to ensure rural aquaculture development.

Setting Programme Goals and Planning

In the light of resource inventory and possible limitations suitable target groups may be identified, programme goals may be set up and accordingly suitable extension strategy may be planned. Without such an early insight and planning, the programme may not have firm and realistic footing. Although the fish farmers are the usual target of any fish culture extension programme, all the fish farmers may not be suitable to be involved for immediate participation.

Target groups may be selected on a number of criteria including farming practice, production level, income, education, cultural background, nature, reputation in the society, initiative, liable to change their attitude, etc. Selection of suitable communication channels is also very important. Data collected during the pre-adoption survey provide the necessary information for such selection. After these selections, programme goals may be set up. Goals indicate the direction towards which the programme is oriented. It also provides reference level for evaluating the programme achievements. Examples of goals in such extension programme may be on the following lines:

- Improving the socio-economic uplift of fish farmers and raising the standard of living;
- Bringing 100 per cent of the available undrainable ponds for composite fish culture.

Programme Implementation

"Plan the work and work the plan" is an appropriate term for any extension programme. Once the plan is laid, all possible efforts should be diverted to ensure that responsibilities are carried out, schedules are followed arid activities accomplished as per the plan. Although the strategies and planning of the aquaculture extension programme are situation specific, some general steps may be cited as follows;

- Through heavy flow of information using mass media, publications, individual and mass contacts, etc., awareness and interest should be created among fish farmers.
- One or two demonstration centres may be set up and the technology of composite fish culture and seed production in undrainable ponds may be demonstrated to maximum possible farmers to let them realize the case of operation, production potential and profitability.
- A set of suitable farmers should be selected initially and be motivated and guided enough so that they strictly follow the different package of practices as per the schedule.
- Proper steps may also be taken to make available the critical material inputs at the pond/farm sites and if the programme permits, subsidy should be given as a token of initial attraction.
- If the availability of quality fish seed is found to be a limiting factor, fish seed may be distributed free of cost or at concessional rate to the farmers at the initial stage. Proper attention may also be paid for extending fish breeding and seed rearing programmes.
- Facilities for proper monitoring of water quality and fish health may be extended through the participation of nearby laboratories.
- Periodical netting for growth check/health inspection should be strictly followed and supervised.
- Self-explanatory/pictorial instruction booklets dealing with basic steps of composite fish culture, control breeding of common carp, techniques of pituitary gland collection, induced breeding of major carps, hatchery operation for carps, nursery and rearing pond managements,

techniques of fish seed transport, etc., may be prepared, explained and distributed among farmers.

- Ad hoc training courses should be organized at the demonstration sites on different aspects of fish culture and fish seed production for participating and other interested farmers. Exhibition programme/Fish Farmer's Day should be organized time to time at different places in which live specimens of all the six carp species, other culturable air-breathing fishes, harmful predatory species, weeds, fish feeds, fish toxicants, etc., should be shown and the objective and goal of the programme may be exhibited through models, charts, posters, etc.
- At times a team of a few farmers may be selected on the basis of their leadership quality and performance and sent to visit important aquaculture centres, farms, research institutions, etc.
- Individual contacts through home visits is a very effective extension method. The extension worker must be very clear in this objective during the visit and must do sufficient preparation with regard to subject matter information he is going to deliver to the fish farmer and family members.
- Evening is the most suitable time for organizing an assembly of farmers. Necessary details about practices to be followed the next day may be explained to them during such assemblies. Teaching aids may be used to make the communication effective.

Programme Evaluation

Programme evaluation is the process to determine the extent of success of the executed extension programme in the light of present objectives. It is an important management function in order to ensure effective implementation of the programme. It also helps in the identification of the deficiencies and weakness of the programme so that proper corrective measures may be taken to make it more useful in its future course. Programme evaluation can be conducted once a year or at a specific period of the programme and finally at the concluding phase of the programme. The process of evaluation also depends upon the nature of the programme. A short term and less extensive localized extension programme may be evaluated by the extension workers themselves through the analysis of progress reports, field records, questionnaires, etc. However, broad-based and elaborate extension programmes can be evaluated by specialists in association with the extension workers to determine the effectiveness and impact of the extension programme.

It is convenient to fractionate the whole programme into smaller components for effective and easy evaluation. Fractionation may be done as follows:

- Resources (financial, personnel an material) made available.
- Objective of the programme in clear terms.

- Phases of the programme (evaluation should also be done phase-wise).
- Data collection from records and tabulation.
- Selection of ways, means and methods for the collection of data/information from participating, non-participating fish farmers, village youths, prominent persons of the locality, etc.
- Sample selection
- Collection of data/information from target and non-target groups
- Tabulation of data
- Data analysis and interpretation of results

To measure the degree of success, certain values have to be associated with the information. Increased fish production level, profit through increased fish yields, knowledge of modern management techniques, fish breeding, fish seed rearing, increased number of ponds/water areas in the area, etc. are some of such measurable values for programme evaluation.

BIOTECHNOLOGY IN FISHERIES DEVELOPMENT

The gap between human demand and the availability of livestock products is huge in Africa, and imports have been increasing fast. The productivity of African breeds is very low: mature beef cattle of 4-5 years hardly weight 300 kg (as against 400 kg at one year for exotic animals); the best African cows produced 300 litre/lactation (as against 5 000 litres and more in Europe); 100 African ewes produced 50 lambs every two or three years (as against 150 lambs a year in developed countries).

Retrospectively, production policies that aimed at developing livestock through cross-breeding of exotic grade cattle with indigenous ones have generally failed. The new approach is not to make selection of local performant animals. Biotechnology offers good methods both for selection and cross-breeding schemes, essentially through Multiovulation Embryo Transfer (MOET) and the field diagnosis of pregnancy. Yet, there is wide diversity among the countries to use new technologies. For instance, even if simpler techniques such as Artificial Insemination (AI) are considered, it is seen that in East Southern Africa the technique is widely used, whereas in West and Central Africa, because of the predominance of nomadic transhumant livestock production, the technique is least developed.

MOET is well developed in few countries such as Zimbabwe and South Africa. Cattle farmers in these countries import frozen embryos from overseas. MOET is promising especially in the development of dairy cattle farms around large cities. Both AI and MOET are being used in Open Nucleus Breeding Systems (ONBS) to improve animal production. Small-scale ONBS projects are being carried out by FAO in small ruminants in Ghana and the Gambia. Moreover, by using semen from bulls, with genetic resistance to high temperature, diseases and insects common to African countries, embryos could be produced from superior cows of other regions and then implanted into surrogate mothers within the environment where improved livestock are needed. In animal feeding, biodegradation of low-quality forage is also a promising technology. It can be associated with chemical degradation techniques (use of urea and ammonia) to improve animal feeds during dry seasons.

In some of the countries, farm diagnosis of pregnancy is also being used by field workers by using a kit based on the difference of levels of progesterone in cow's milk.

In the field of animal health, Sterile Insect Technology (SIT) is well known in Africa. Sterile males of glossina are being reared in Bobo-Dioulasso and Burkina Faso for field application. Successful field projects in tsetse control have been carried out in Nigeria (BICOT Project) and in Burkina Faso. Newworld screwworm was eradicated in the Libyan Arab Jamahiriya in 199091 by the use of SIT.

In disease diagnosis, ELISA is being used to evaluate rinderpest immunity in vaccinated animals. Monoclonal Antibody Techniques (MAB) is not commonly used. The two techniques are very promising because they are very simple and most effective. Trials are under way to use the ELISA test in trypanosomiasis diagnosis. In vaccine production recombinant rinderpest vaccine trials done in the United States seem ready to be utilized in a mass vaccination campaign against rinderpest in Africa. There is an increasing level of activity in hybridoma work in sub-Saharan countries.

The work of ILRAD in Kenya is addressing the problem of diseases and parasites that affect cattle. Some of the common cattle diseases are East Coast Fever, tickborne diseases and trypanosomiasis. The focused approaches to researching on these diseases is expanding to yield results that will significantly benefit cattle breeders. The development of rinderpest vaccines remaining stable at ordinary temperatures was of high practical value.

Biotechnology in fisheries is recent in Africa and it is being applied essentially in private fisheries sector. Work is mostly confined to hormonal treatments for sex reversal and pawning, detection of fish and shrimps disease and development of fish feed.

COMMERCIALIZATION AND ROLE OF THE PRIVATE SECTOR

In general, commercialization of agricultural biotechnology is far behind the application of biotechnology in human health and medicine. In the developed countries, especially in Japan, the private sector plays a major role in the generation, development and commercialization of biotechnology products and techniques. Because of commercial considerations, concentration has been mainly on diagnostics, vaccines, pharmaceuticals and other health-related products. Moreover, because of the eminent involvement of the private sector, marketability of the products and potential return on investment are crucial factors in deciding which products are to be developed and commercialized. In Japan, the MAFF is subsidizing fundamental technical development projects in the private sector and providing funds for investment in such projects. It is also offering financial assistance to prefectural governments so as to encourage their R&D activities and utilization of the outcome of such activities. The Ministry takes steps to subsidize the research projects of private technical research associations and other similar study programmes. These subsidies are offered mainly through the Ministry's project for the development of biotechnology and other high technology for the food industry. In fiscal 1989, MAFF began such financial assistance programmes as the "Transformation of plant cells through the introduction of small organelles into the cells" and "Genetically manipulated vaccines against animal protozoa diseases".

In the area of biotechnological R&D activities in the food industry, the Ministry started the project "Development of bioreactor systems for the food industry" in 1984. This project has achieved a number of remarkable results regarding the conversion of saccharides, fat and other substances. Another project, the "Development of technology for improvement of enzyme functions for the food industry" aims to improve the functions of enzymes for food production utilizing recombinant DNA technology. In fiscal 1989, the Ministry began a new programme, "Food production by large-scale, high-density fermentation under ultra-high pressure conditions" with a view to increasing the efficiency of the food production process under ultra-high pressures and to finding food materials with a higher value added.

Two more projects were launched in 1989: the "Utilization of extracts from trees (green sprouts project)" and "Advanced and multipurpose utilization of unused marine resources (marine-frontier project)". The objective of these projects is to develop the technique for extracting useful substances from trees and marine organisms and utilizing them for foods, perfumes and medicines, thereby helping further growth of forestry and fishery industries.

The Bio-Oriented Technology Research Advancement Institution (BRAIN) was founded in October 1986 by the joint investment of the government and the private sector. In an effort to encourage the R&D activities of the industries related to agriculture, forestry, fishery and food production, this institution is investing in joint technical development corporations and is providing no interest-bearing loans to the technical research projects of businesses on certain conditions.

During three years from fiscal 1986 to 1988, BRAIN invested in 14 study projects and offered loans to 70 projects. In fiscal 1989, the Institution invested in, among others, the Institute of Aquacultural Technology of Cold Water Fishes, Wakayama Agro-biological Research Centre and the Japan Turf Grass Co. The first is engaged in the development of aquaculture systems of cold water high-class fishes, the second in R&D activities for rationalized process of tangerine juice production and the third in the breeding of better turfs. In the area of biotechnology-related R&D in the private sector, financing systems through the Japan Development Bank, the Agriculture, Forestry and Fishery Finance Corporation and other similar government organizations were reinforced to turn the results of these R&D activities to practical use. It is hoped that as biotechnology is increasingly introduced to industries, these financial measures will be utilized more in the coming years.

Prefectures are becoming more and more interested in R&D activities on biotechnology and the practical use of the outcome of such activities. This is evident from the fact that almost all prefectures throughout Japan have a council on this technology composed of scholars and other people of expertise.

MAFF is offering to prefectures a variety of subsidies in order to diffuse the mass production technology of virus-free seedlings of vegetables, ornamental plants and fruit trees and the transplanting technology of the embryos of beef cattle. In fiscal 1989, the Ministry started the "Project for establishing embryo supply centres" to ensure a stable supply of embryos of beef cattle. This project is improving the raising and controlling facilities of cattle from which such embryos are taken.

Since 1986, MAFF has also been implementing the "Project for assisting local R&D activities in biotechnology". This project offers financial assistance to the R&D activities of prefectural research institutes concerning the breeding and utilization technology of biological resources. Other biotechnology-related activities of the Ministry include regional biotechnology meetings held in seven locations across the country with the object of exchanging information on R&D. The Ministry's new initiative in fiscal 1989 included surveys and studies for examining the direction of the practical use of biotechnology in each region. The Tsukuba Bioscience Hall created in 1989 provides a forum for collaboration among trainees from industries, universities and the government and international exchange programmes. Furthermore, to foster links among genetic resources and biotechnology, the Ministry has a strong national genebank and germplasm conservation programme.

Of the developed countries in the region, the involvement of the private sector in biotechnology, especially plant biotechnology, is rather limited in New Zealand. Very little research on plant biotechnology is being carried out by private companies. A shortage of venture capital, low taxation incentives and tough competition from large overseas firms make such ventures unattractive.

Several small companies are engaged in either product formulation, or manufacture of biological control agents. Larger international companies (*e.g.* Monsanto, MPI Koln, PGS Ghent, ICI) and overseas research institutions (UCLA, California; John Innes Institute, United Kingdom; Christian Albrecht University, Germany) have provided personnel, facilities and some supportive funding on collaborative projects. Producer Boards (*e.g.* New Zealand Kiwifruit

Marketing Board) also support research through joint funding of government programmes.

Many New Zealand research programmes on plant biotechnology are not mature enough to have reached the stage of successful commercialization. In addition, some programmes are commercially sensitive, particularly where research is collaborative with private companies. Examples of plant biotechnology research that have been, or are in the process of being commercialized in New Zealand include:

- Establishment of a commercial tissue culture laboratory for clonal Forestry with Radiata Pine (FRI).
- Use of immuno-molecular techniques for developing test assay kits and for analysis of gene expression (DSIR, Fruit and Trees).
- Plant variety rights have been granted for four new plant varieties.

Future commercialization is likely to be through the release of improved crop varieties, development of new cultivars and the licensing of technologies that have been developed.

In the developing countries, in vitro culture for micropropagation of elite and disease-free materials, production and distribution of efficient nitrogen-fixing microbes, diagnostic kits and monoclonal antibody-based vaccines are the main areas that have been commercialized to varying degrees in different countries.

In developing countries, the role of the private sector in modern biotechnology has been rather limited. This is attributed partly to: (i) vague government policies regarding the private sector; (ii) unclear policies or no policies/views on patents and intellectual property rights; (iii) poor links between public and private sectors; (iv) low purchasing power for new products that are usually highly priced and are out of reach of the majority of resource-poor farmers and low-income consumers; and (v) inadequate local expertise and infrastructure and R&D support to new biotechnology.

In India, biotechnology industries have recently been growing in the private sector. Hindustan Lever Ltd, an Indian subsidiary of the multinational Unilever, the Tata Energy Research Institute (TERI) and Southern Petrochemical Industries are some of the leading biotechnology companies in the country. These have considerable in-house R&D bases active in the application of new technology to agribusiness, such as enzyme technology for high-value chemicals and quality edible oil by using genetically modified bacteria; protoplast fusion in yeast for efficient fermentation and production of edible oil from biomass; tissue culture of plantation and ornamental crops; biological nitrogen fixation; organic components for higher rate of photosynthesis; big-insecticides; plant growth promoters; biomass processing for animal feeds; and hybrid seeds. Some small and medium-size companies are also initiating biotechnology activities but, with less R&D capacity, are more wary of what they consider to be a higher-risk technology. The private sector in India is particularly active in tissue culture products. There are about 50 private sector companies such as Indo-American Hybrid Seeds, A.V. Thomas and Co., Unicorn Biotek, the three companies and others that are comprehensively involved in commercial in vitro production of planting materials of several horticultural and plantation crops such as cardamom, coffee, tea, oilpalm, orchids, strawberry, roses, Spatiphyllum, banana, lily, Gerbera, etc. both for domestic and export markets. These companies have established effective links with the public sector.

The recent liberalization of trade, including that of certified seed, coupled with the greater priority given to science and technology and the promotion of private sector and public-private sector links, should encourage increased participation of the industry in agricultural biotechnology R&D. Financial incentives are available for supporting the growth of indigenous biotech industries. However, venture capital is still lacking. Several multinationals, *e.g.* Seedtec International, Cargil Inc., Dehlgien Inc., Northrup King and Sandoz (through their local subsidiaries) have been shortlisted by the Indian Government for production of hybrid seeds which may involve biotechnologically produced seeds in collaboration with Indian companies. To take advantage of the availability of expert personnel in biotechnology at relatively low costs, a few multinational companies like Astra AB (Sweden) are setting up a biotechnology R&D centre in India. Recently, Pro-Agro, an Indian seed company, established links with Plant Genetic Systems, a leading Belgium-based plant biotechnology company, for the production and distribution of hybrid seeds.

In Indonesia, there is relatively little direct involvement or sponsorship of research in government institutions at present. Some companies are investigating improved planting materials. A commercial production system has been started for improved oil palm plantlets, with a target of 1 million plants per year in 1993, at the Research Institute for Oil Palm at Marihat. Animal vaccines are produced by Pus Vetima and human vaccines by Perum BioFarma, in government-owned companies, which also undertake some biotechnology research.

In the Republic of Korea, the entrepreneurship drive led the biotechnology sector to be organized in 1982 under the Korean Genetic Engineering Research Association (KOGERA). At present more than 20 private companies are participating in this Association. The government provides the members of KOGERA with subsidy as the seed money. In 1980, the public sector provided 70 per cent of the R&D support to the biotech industry. Six years later, the situation was reversed; in 1986 the public sector provided only 28 per cent of the R&D cost whereas the industry provided 72 per cent of the cost. This shows the increasing success and confidence of the private sector. The private companies are mostly involved in fermentation and pharmaceutical matters and have very little interest in agricultural biotechnology. Most of the companies are using biotechnology for the production of antibiotics, insulin, new vaccines and interferons.

The Association performs an important role in the exchange of biotechnology information and also promotes relevant research. MOST provided a special fund for 122 projects in the field of gene manipulation and recombinant DNA technology in 1990. Of these national projects, 32 projects were directly connected with crop improvement. The Korean Science Foundation (KSF) has contracted research projects to universities relating to biotechnology in the fields of medical, pharmaceutical, and agricultural sciences.

The entire biotechnology work in agriculture in the Republic of Korea is coordinated by the Agricultural Biotechnology Research Council comprising researchers from national institutes and universities and representatives of industries and the private sector. A few private companies involved in agricultural biotechnology are using the new technology for the production of virus-free potato planting material, and through the anther culture technique are producing haploids for heterosis breeding in Chinese cabbage and rice. The private sector, besides having strengthened its own R&D system, also provides assistance to universities for undertaking specific biotechnological research. The increase in personnel and funds for biotechnological research and development in the private sector in the Republic of Korea.

INTELLECTUAL PROPERTY RIGHTS

Two Intellectual Property Rights (IPR) mechanisms namely, patents and plant breeders' rights are most important for agricultural biotechnology. A patent is a right granted by the government to inventors to exclude others from imitating, manufacturing, using or selling a specific invention for commercial use during a certain period, usually 17-20 years.

The patent holder, in turn, is obliged to disclose the invention to the public. Plant Breeders' Rights (PBR) are rights granted by the government to plant breeders to exclude others from producing or commercializing material of a specific plant variety for a period of about 15 to 20 years.

Both under IPR and PBR, use of the protected matter is restrictive, being highly restrictive under IPR. There is, however, provision for research exemption under IPR which allows others to study the protected subject-matter without reproducing or multiplying it for commercial purposes.

Under PBR, there are provisions for breeders' exemption and farmers' privilege. But, in 1991, UPOV tightened PBR by eliminating the breeders' exemption for an essentially derived variety-a variety predominantly derived from another (initial) variety which retains the expression of the essential characteristics from the genotype of the initial variety. Now, as per the 1991 legislation, a breeder who inserts a single new disease-resistance gene into a

PBR-protected variety will have to obtain permission from the holder of the original rights before marketing the new variety. In 1991 UPOV also attempted to tighten up farmers' privilege, *i.e.* freedom to farmers' freedom to use their own harvested material of protected varieties for the next crop on their farm, but because of a lack of consensus among UPOV members it was left to the national governments whether to permit farmers to reuse the seed of a BPR-protected variety for further crop cycles on their own holdings.

The status of development and adoption of IPR and PBR varies widely from country to country in the region. The three developed countries (Australia, Japan and New Zealand) are all members of UPOV and have explicit IPR provisions for biotechnological products and procedures. None of the developing countries are members of UPOV and have mostly been opposed to enforcement of any form of IPR, particularly patents.

However, lately, under various kinds of bilateral pressures, such as the special 301 Provisions of the US Omnibus Trade Act of 1988, as well as multilateral pressures, and after having signed the successfully negotiated GATT agreements in December 1993, whether to introduce legislation covering proprietary protection for biotechnology products and procedures, and/or how rigid the protection system should be, are under active debate in several developing countries of the region.

Those in favour of having the legislation argue that an effective IPR system would: (i) improve the access of the country to new technology from abroad; (ii) stimulate private sectors; and (iii) encourage foreign investment.

Those against it argue that in the context of the developing countries it will: (i) suppress in-country innovation and local industries and production systems; (ii) increase the dependence on foreign countries even for materials and procedures that are vital for the people of the importing country; (iii) increase prices of even commonly used products and services; and (iv) restrict information flow and knowledge growth.

The effects of IPR on the development of biotechnology in developing countries are not clearly understood and should be studied systematically, covering countries at various levels of development and of varying sizes and in the context of different industries, such as pharmaceuticals, agriculture, food industries, environmental protection, etc.

Thailand adopted a new patent act (1992) that protects biotechnology inventions after the United States Government had withdrawn Thailand's trade benefits under the General System of Preferences. Under a similar pressure, when the renewal of the Sino-American Science and Technology Agreement was postponed by the United States Government, China revised its patent law effective as of January 1993, which protects biotechnology inventions, but excludes plant and animal varieties. In India, the matter is hotly debated. Despite the pressure from the United States Government by postponing the Indo-US

Vaccine Action Programme and the Indo-US Science and Technology Initiative, the Indian Government has not yet strengthened its IPR legislation. However, given these pressures, the changed and liberated seed policy of the government and increasing pressure from the private seed companies within the country, there is a change in the mood of the government, although strong opposition persists. India signed the GATT agreement in December 1993. Its 1994 Spring Session of the Parliament held a special debate on the issue, and the opposition parties continue to agitate. It appears that a consensus may emerge to evolve an IPR system that will not promote inequity, jobless growth, and ecological degradation. The government is currently preparing an effective sui generis UPOV PBR, explicitly emphasizing breeders' exemption and a strong farmers' privilege. In addition, the system would provide for farmers' rights and include mechanisms for the operation of these rights.

As regards other developing Asian countries with considerable programmes in agricultural biotechnology, no patent protection systems exist either in Malaysia or the Philippines. However, Indonesia is preparing a new IPR legislation.

It can be seen from the above that, under various kinds of pressures, from outside or within, and as biotechnological R&D efforts are being intensified, several developing countries are moving towards restructuring or strengthening their IPR systems.

What is important at this stage is that the countries strike a rational balance between the positive and negative aspects of the IPR system and tailor their legislations to their needs, opportunities and aspirations. In countries where the local technology base is weak, an IPR system may contribute little to the indigenous technology capacity.

However, if the country is now able to pursue cutting-edge biotechnology and produce innovative and useful products and processes at competitive prices, the relevance of an appropriate IPR would increases. The countries must therefore be able to analyse the various pros and cons and short-and long-term socio-economic and ecoenvironmental advantages and disadvantages of instituting a particular IPR system. International organizations, such as FAO, should assist the countries in this task.

REGULATORY ASPECTS

Biosafety

R&D biotechnology is heavily influenced by two regulatory issues, namely biosafety and intellectual property rights. Biosafety aspects include the policies and procedures needed to ensure the environmentally safe development and application of biotechnologies. An effective biosafety system to regulate release and use of Genetically Modified Organisms (GMOs) and to address concerns about potential risks to public health and environment must be in place in all countries involved in biotechnology. Biosafety regulations are necessary also to facilitate technology transfer. Comprehensive biosafety guidelines and measures exist in the developed countries, but the situation is far from satisfactory in most developing countries, although they are aware of the need and are in the process of formulating national frameworks, procedures and capabilities for addressing biosafety issues.

In Australia, biosafety guidelines/regulations have been evolving since 1975, when the Academy of Science Committee on Recombinant DNA (ASCORD) issued its first guidelines. In 1981, as per the Commonwealth resolution, the Recombinant DNA Monitoring Committee (RDMC) was established. In 1987, the RDMC published "Procedures for the Assessment of the Planned Release of Recombinant DNA Organisms". The RDMC was further transformed into the Genetic Manipulation Advisory Committee (GMAC) to include the surveillance of any genetic manipulation procedure resulting in an organism containing foreign DNA that was unlikely to be formed by conventional breeding practices. The GMAC emphasizes that the focus of the assessment should be directed towards ensuring that the release does not cause harm to people, livestock, plants or the natural environment, and not primarily be directed towards considering how the organism was constructed.

GMAC adopted the following procedure:

1. Approval of the concerned Institutional Biosafety Committee (IBC) is obtained.
2. The IBC sends the proposal to GMAC.
3. GMAC, according to its Procedures for the Assignment of the Planned Release of Recombinant DNA Organisms, in consultation with its Scientific and Planned Release Committees, analyses and evaluates the proposal, especially the genetic aspects and any possible environmental impact associated with the construct.
4. The above assessment report, especially pointing out hazards, if any, is sent to the proposer and to the agency that has legal jurisdiction over the release of that particular class of organism.
5. The responsible agency then makes the decision whether to release or not to release the organism based on the report from GMAC and other related considerations."

In Japan, five guidelines have been established for the regulation of research of DNA organisms and the safe promotion of experiments since 1979. Appropriate regulations are considered effective in minimizing the potential risks of a new technology to the environment and human health, and in ensuring positive public perception. The purpose of MAFF's "Guidelines for the Application of Recombinant DNA Organisms in Agriculture, Forestry,

Fisheries and the Food Industry and Other Related Industries", is to promote the safe progress of agro-industries by: (i) defining general principles for the appropriate application of rDNA organisms; and (ii) ensuring safety in the use of rDNA organisms. The guidelines are based on OECD guidelines.

Applications for field trials of rDNA organisms are to be approved by MAFF through biosafety assessments conducted by the authority committee under MAFF. The first application for field tests using transgenic tomato plants harbouring TMV coat protein genes was filed in late 1990. The application was based on three years of greenhouse experimentation. As the biosafety review of this application was completed, the first field trials were approved in 1991 and since then many field tests have been approved.

In New Zealand, an Advisory Committee on Novel Genetic Techniques (ACNGT) was set up in 1978 under the terms of a Cabinet resolution. All persons working in the public sector (universities, government departments, etc.) are required to inform ACNGT of their intention to conduct experiments that are defined as "any experiment involving novel genetic techniques that is properly contained and that can be terminated at any point without loss of containment".

In 1988 a Genetically Modified Organism Interim Assessment Group (JAG) was established under the Environment Act. The principal role of IAG is to assess all proposals to field test or release GMOs and in so doing provide effective opportunity for public involvement in the assessment process. Private sector researchers are not bound by IAG but are encouraged to use it. In 1989 the New Zealand Government announced that a new agency, the Hazard Control Commission, would be established. This Commission, in addition to responsibilities with respect to hazardous substances, is responsible for assessing and licensing all genetic manipulation work in New Zealand and approving applications to import or release new organisms.

Among the developing countries, comprehensive national biosafety systems/guidelines have been set up in China, India, the Philippines and the Republic of Korea. Most of these guidelines are based on multitier approach, modelled on the lines of OECD and NIH and NAS (United States) biosafety procedures and guidelines for handling and release of GMOs.

In India, the Recombinant DNA Advisory Committee of the Department of Biotechnology (DBT, 1990) prepared a set of Recombinant DNA Safety Guidelines that cover all areas of research and large-scale operations involving genetically engineered organisms, excluding humans. The institutional mechanism for implementing the guidelines consists of:

- The Recombinant DNA Advisory Committee, under DBT, to formulate and update biosafety guidelines;
- Institutional biosafety committees located at all centres engaged in genetic engineering research and production activities;
- The Review Committee on Genetic Manipulation, under DBT, to guide the institutional biosafety committees;

- The Genetic Engineering Approval Committee, functioning under the Department of

Environment, to review and approve activities involving the large-scale use of genetically engineered organisms and their products in research and development, industrial production, environmental release, and field applications.

In the Philippines, the National Committee on Biosafety and Biosafety Guidelines is a working group composed of representatives of the Department of Agriculture, the Department of Science and Technology, the University of the Philippines, and the International Rice Research Institute.

It has prepared biosafety guidelines that include a provision for case-by-case review of proposed releases of genetically modified organisms. A national Biosafety Committee was established in 1990 to oversee the compliance of policies and guidelines in public and private institutions. A network of 39 Institutional Biosafety Committees was created to ensure the implementation of the guidelines at the institutional level. The National Biosafety Committee coordinates with other national agencies involved in regulations, such as the quarantine services and the environmental management bureau. To date, the National Biosafety Committee has approved 15 applications for the importation of, among other things, transgenic cotton tissue and transgenic strains of rice.

GENETIC AQUACULTURE RESOURCES

Genetic resources are the foundation on which species, stocks and genetically-improved strains are based. At the species level, more aquatic animals are being farmed now than ever before.

Although the common carp, Cyprinus carpio and goldfish, Carasius auratus, were domesticated several thousand years ago into a variety of shapes and colours, most of the farmed fish today are very similar to their wild relatives. Improvements in our knowledge of artificial reproduction, reproductive biology, early larval rearing (training series) and basic genetics have recently allowed fish breeders to improve genetically species such as rainbow trout, coho and Atlantic salmon, channel catfish, Nile tilapia, as well as common carp.

The culture of several important species still relies on the collection of brood stock or seed from natural populations. Perhaps the most important group of species whose culture is dependent on natural populations is the marine shrimp Penaeus spp. Shrimp farming in South and Central America stock production ponds with wild-caught larvae. Hatcheries that produce shrimp larvae exist in Asia and the Americas, but the broodstock are generally collected from the wild. Culturists recognise the problems associated with this harvest of wild resources and are taking steps to domesticate marine shrimp. Other culture systems dependent on wild resources include milkfish in the Philippines, yellow tail in Japan, and eel in Asia and Europe.

However, simply having a domesticated species or genetically improved species is not sufficient to guarantee optimum production from an aquaculture facility. In addition to proper husbandry, *i.e.*, water quality, nutrition, health etc., broodstock must be managed to ensure conservation of genetic resources, to maintain the desirable characters of the farmed species and to avoid problems of inbreeding.

CURRENT STATUS OF AQUACULTURE GENETICS

DOMESTICATION AND STRAIN EVALUATION

When wild fish are moved to aquaculture settings, a new set of selective pressures comes into play that often changes gene frequencies. Thus an organism better suited for the aquaculture environment begins to evolve. This process, termed domestication, occurs even without directed selection by the fish culturist. Domestication effects can be observed in some fish within as few as one to two generations after removal from the natural environment. In channel catfish (Ictalurus punctatus) an increased growth rate of three to six per cent per generation was observed.

The oldest domesticated strain of channel catfish (89 years), the Kansas strain, has the fastest growth rate of all strains of channel catfish. Although most domesticated strains usually perform better in the aquaculture environment than wild strains, there are exceptions, *e.g.*, wild Nile tilapia, Oreochromis niloticus, and rohu, Labeo rohita, grow better in the aquaculture environment.

The explanation for this appears related to a lack of maintenance of genetic quality and genetic degradation in domesticated strains. Poor performance of some domestic tilapia is related to poor founding (parental) lines, random genetic drift, inbreeding and introgression with slower growing species, such as O. mossambicus and slower growing strains such as Nile tilapia from Ghana. Channel catfish strains differ in growth, disease resistance, body conformation, dressing percentage, vulnerability to angling and seining, age of maturity, time of spawning, fecundity and egg size. Strains of rainbow trout, Oncorhynchus mykiss, show similar variability. Domestication of farmed shrimp (penaeids) has been relatively slow compared to that of finfish.

This can be attributed to use of wild broodstock and postlarvae, a lack of understanding of shrimp reproductive biology for domestication of the species and perceptions of low potential for genetic improvement. Current reliance on wild broodstock is risky and negates the opportunity to enhance disease resistance (as well as other production traits) through selective breeding. Efforts to domesticate broodstock are now hampered by endemic disease challenges; however, recent collaborative research between the Commonwealth Scientific and Industrial Research Organisation (CSIRO) in Australia and the shrimp culture industry has resulted in successful captive breeding of Penaeus japonicus. Economic analysis has demonstrated that domesticated broodstock are more cost-effective than wild broodstock and that reproductive performance of domesticated P. monodon can match that of wild broodstock of a similar sise.

Use of established, high-performance domestic strains is the first step in applying genetic principles to improved aquaculture mana-gement. Strain variation is also important, since there is a strain effect on other genetic enhancement approaches, such as intraspecific crossbreeding, interspecific hybridisation, sex control and genetic engineering.

INBREEDING AND MAINTENANCE OF GENETIC QUALITY

It is as important to prevent production losses due to inbreeding as it is to increase production from genetic enhancement. This applies especially to species with high fecundity, *e.g.* Indian and Chinese carps, where few broodstock are necessary to meet demands for fry and broodstock replacement. The detrimental effects of inbreeding are well documented and can result in decreases of 30 per cent or greater in growth production, survival and reproduction.

INTRASPECIFIC CROSSBREEDING

Intraspecific crossbreeding (crossing of different strains) may increase growth rate but heterosis (differences between offspring and parents) may not be obtained in every case. Increases of 55 per cent and 22 per cent in growth rate of channel catfish and rainbow trout crossbreeds, respectively, were achieved using this technique. Chum salmon crossbreeds, however, have shown no increase in growth rates compared with parent strains.

Common carp crossbreeds generally express low levels of heterosis, however, those that exhibited positive heterosis are now the basis for carp aquaculture in Israel, Vietnam, China and Hungary.

The crossing of common carp lines in Szarvas, Hungary is a good example of the relative success of crossbreeding. During the last 35 years, more than 140 crosses have been tested.

Three were chosen, based on ~20 per cent improvement in growth rate (and other qualitative features), compared to parent and control carp lines, for culture purposes. Now approximately 80 per cent of common carp production comes from these "Szarvas" crossbreeds. Production of gynogenetic female lines and gynogenetic sex-reversed inbred male lines from common carp with the best combining ability was an important part of the Hungarian crossbreeding programme. A higher heterosis was expected from crossing inbred lines, but the growth rate of F1 crossbreeds was only 10 per cent higher than controls.

Kirpichnikov successfully produced a new strain of cold resistant carp - the "Ropsha" carp - for cold zones in northern Russia using local carps and

Siberian wild carps from the River Amur. The Czech Republic also had a national common carp breeding programme and improved growth rates with crossing, "South Bohemian" x "Northern mirror" carp and the "Hungarian 15" x "Northern mirror". Likewise in Israel, over 20 years of crossing common carp strains revealed that the cross using the strain "DOR-70" and the Croatian line "Nawice" produced fast growth, and this is now one of the most popular crosses for Israeli carp production.

In Indonesia, strain development using artificial gynogenesis and sex-reversal resulted in 10 common carp inbred lines, which were later used for crossbreeding. In Vietnam, eight local varieties of common carp, along with "Hungary", "Ukraine", "Indonesia" and "Czech" strains are maintained, with significant heterosis observed in F1 generations of crossbreeds.

Double crosses among Vietnamese, Hungarian and Indonesian strains have subsequently been used for carp selection and crossbreeding programmes throughout Vietnam.

The Vietnamese x Hungarian common carp crossbreed is particularly popular, due to fast growth and high survival rates under different production conditions. Under various rice-field conditions, growth rates of different strains of Nile tilapia and their crosses showed that the crosses were superior to pure Senegal strains.

Breeding programmes are also under development in several countries for the Java or silver barb, Barbodes (formerly Puntius) gonionotus, another economically important fish species in Southeast Asia. The Bangladesh programme used three strains: "Bangladesh", "Thailand" and "Indonesian".

The growth rate of females from six crosses was 23 per cent higher than the average growth rate of the parent strains. Even higher growth rates (35 per cent improvement) were found in the three crosses using the Thailand strain as either the sire or dam. In the Vietnamese breeding programme, six different strains were used to produce a population ideally suited for culture.

Cross-breeds of different strains of European catfish, Silurus glanis, are characterised by outstanding adaptability under warmwater holding conditions and mixed diet feeding regimes.

In addition, crosses with the walking catfish, Clarias macrocephalus, have shown improved resistance to Aeromonas hydrophila infections. Studies on "domestic a" × "domestic b" channel catfish also showed greater heterotic growth rates than "domestic" × "wild" crosses (Dunham and Smitherman, 1983. The same was found in rainbow trout crosses. Strain mating incompatibilities, however, can occur and can impede fry output.

APPROACHES THE GENETICS OF BEHAVIOUR

We describe in order of increasing complexity how a researcher might approach the genetics of a behaviour of interest. For each method, I describe its aim, the general approach, its strength and weaknesses and the kinds of traits and animals for which it is most suited.

'Low tech' approaches

The first step in approaching the genetics of a behavioural trait is to determine if the behaviour is repeatable, which involves measuring the same behaviour on the same individual several times. Repeatability is the proportion of phenotypic variance attributable to the individual, which could be caused by additive genetic variance or environmental variance with long-lasting effects. Evidence that a behaviour is repeatable indicates that it might have a heritable component and is amenable for further genetic dissection.

The repeatability of a trait sets an 'upper bound' to heritability. However, it is worth considering that non-genetic effects can also produce stable behaviour; repeated reinforcement including learning can also lead to stability. Another indication that there might be a heritable component to the behaviour is if it is consistent across contexts; in this case, the stability occurs via a correlation between individual behaviours in different contexts, or a behavioural syndrome.

Another relatively low-tech approach to the genetics of a behaviour is to ask if it differs across populations that occur in different kinds of environments. If so, then population differences in the behaviour could reflect an evolved response to differing selective pressures. Alternatively, the differences could reflect environmentally-induced changes within individuals.

A way to disentangle these two hypotheses is to rear animals from the different populations in a 'common garden', or in the same environment. If population differences in the behaviour are preserved under common environmental conditions, then the difference between populations might be genetic in origin.

A common garden experiment represents the first step towards approaching the genetics of complex traits such as behaviour, but there are at least two important factors to consider before concluding that population differences are genetic in origin. The first is that F1s reared in a common garden could still experience parental effects from their wild-caught parents. To control for such environmental effects, it is preferable to rear the offspring of lab-born individuals and to compare the phenotype of the F2s. Second, GxE interactions should also be considered when populations do not differ in a common environment. For example, individuals from both populations might converge in their response to the common garden environment, but the extent of response might have a genetic component. A reciprocal transplant experiment might reveal genetic differences that are not apparent in a common environment.

An advantage to using population differences as a clue to which behaviours are likely to be heritable is that traits that differ across populations in different environments are probably linked to fitness. Several common garden studies on fishes have shown that population differences in behaviour are often preserved in a common garden. For example, Lahti *et al.* 2001 found evidence for a genetic basis to aggressive behaviour in brown trout by comparing different types of populations. The authors reared fish from 10 populations in a common environment. The populations can be grouped into the following types: resident (non-migratory), sea-run (migratory) or lake-run (migratory), so there were replicate populations for each type. Contrary to the expectation that resident trout are more aggressive than migratory forms, they found that sea-run populations were consistently more overtly aggressive towards opponents than the other types of populations.

There is a rich literature comparing domesticated and wild populations of salmonids, which has revealed genetic influences on several types of behaviour (reviewed in). When reared in a common garden, hatchery fish show reduced antipredator responses are more bold towards a novel object and are frequently more aggressive. However, wild Atlantic salmon (*Salmo salar*) became dominant over hatchery salmon if they were given an opportunity to establish residence, suggesting that the expression of genetic differences between hatchery and wild salmon depends on the environmental context.

Quantitative Genetic Approaches

There is a long history of examining the genetic basis of continuously distributed traits using quantitative genetic techniques, which are based on the phenotypic resemblance among relatives due to shared genes. Unlike some of the reverse methods, quantitative genetic techniques measure phenotypes, not genes, and rely on the resemblance among relatives to infer genetic effects. Quantitative genetic approaches assume that many genes of small effect influence the phenotype, thereby producing a continuous distribution (however, recent QTL studies have called this assumption into question). By measuring the phenotype on individuals of known relatedness, we can partition the phenotypic variation into environmental, genetic and gene x environment variance components.

The simplest quantitative genetic approach is to ask whether the trait of interest runs in families, which would suggest that there might be a genetic component to it. However, families often share a common environment so cross fostering experiments are useful here. Like repeatability, a demonstration of full-sib resemblance sets an upper limit to the heritability. More complicated breeding designs involve estimating the resemblance of parents and offspring using regression, or generating full and half-sib families to disentangle parental effects from genetic effects using ANOVA-based approaches, which are thoroughly described elsewhere. These approaches estimate heritabilities, genetic correlations, parental effects and non-additive genetic variance.

The most powerful quantitative genetic experiments estimate the G matrix. The G matrix refers to the multivariate matrix of genetic variances and covariances between several different traits. Although laborious to quantify, obtaining a G matrix allows us to predict the consequences of selection on any given trait, and, with a few assumptions about the stability of the G matrix through time, allows us to retrospectively analyze selection in the past. In addition to satisfying the assumption of stability required for retrospective selection analysis, it is also desirable to compare G matrices to determine whether genetic constraints are limiting. If genetic correlations can be uncoupled, then the relationship between traits might respond to selection, and the G matrix will, itself, evolve. Alternatively, genetic correlations might limit the number of different configuration of traits that are possible, which would be reflected in an invariant G matrix through time. Fortunately, there are several different techniques available to compare the structure of G matrices but there is debate about the best method.

A relatively new powerful technique known as the animal model allows us to estimate genetic variances and covariances without performing breeding experiments in the lab. Therefore this method is particularly useful for field studies. The animal model is powerful because it takes advantage of all types of resemblance between relatives to partition variance components; it uses all the available information about relatedness from a pedigree. Moreover, this method is suitable for unbalanced datasets, and can accommodate missing values.

Another approach for estimating genetic variances and covariances is to perform an artificial selection experiment, which involves selective breeding of individuals.

Selection experiments have shown that many behavioural traits can respond to selection. Artificial selection experiments are only suitable for organisms that can be kept in the lab, and which ideally have a short generation time. Selection experiments should only be undertaken when different selected lines can be replicated and when sample sizes are sufficient to control for drift and inbreeding. The pitfalls and perils of selection experiments are reviewed in.

Artificial selection experiments can be particularly insightful when they vary the environmental context in which selection occurs. For example, in a very interesting series of selection experiments on medaka (*Oryzias latipes*, Ruzzante and Doyle (1991, 1993) showed that the correlated response of aggressiveness to selection for fast growth depended on the ecological context in which the selection took place. In these experiments, Ruzzante and Doyle selected for fast growth under two different conditions: when food was clumped and could be defended, and when food was dispersed. The same limited amount of food was added to each treatment. Selection for fast growth produced a correlated response to selection on aggressiveness, but only when food was defensible. Interestingly, levels of aggressiveness *decreased* in the fast-growth lines, which they interpret as reflecting indirect selection for 'social tolerance'.

One of the strengths of a selection experiment is that it can tell us whether there are packages of traits that are linked together and respond to selection in concert. For example, selection for increased and decreased stress responsiveness, as measured by the change in circulating concentrations of cortisol in response to handling stress, produced several correlated effects on physiological and behavioural measures in rainbow trout. Relative to trout that did not release very much cortisol in response to handling stress, 'high responding' trout were more aggressive in a new environment were more active in the presence of an intruder and took longer to acclimate to a new environment. Moreover, these behavioural differences were also accompanied by changes in the brain monoaminergic systems. These results suggest that an entire suite of physiological and behavioural changes accompanied modifications to stress responsiveness. Interestingly, the packages of physiological and behavioural traits that changed together are analogous to different 'coping styles' that have been identified in other vertebrates.

Candidate Gene Approaches

Over the past few decades, there has been increasing evidence that the molecular functions of many genes are highly conserved across species. For example, studies on transgenics have revealed that genes from one species accomplish similar functions in distantly-related species. The incredible conservation of gene function in living things allows us to apply genetic information about other species to the organism of interest. This means that we can apply the genetic information gained from studies on model organisms to non-traditional models for which genomic information is not available.

Genes could become candidates based on genetic studies in other species (*i.e.* when polymorphism has already been identified and associated with behavioural variation, Fitzpatrick *et al.*, 2005), or because a physiological pathway leading to behaviour of interest is well-understood, suggesting hypotheses about which genes to examine. For example, the neuroendocrine mechanisms underlying circadian rhythms are well described for several mammalian species. Therefore, we can ask whether the expression or structure of genes along those pathways differ among individuals or among groups. Candidate gene expression approaches typically involve measuring the abundance of mRNA transcript in a particular tissue (usually brain, in the case of behaviour) using quantitative real-time PCR or Northern blots.

One of the attractions of the candidate gene approach is that detailed genomic information on the species is not necessary, therefore this technique can be applied to non-model organisms by designing degenerate primers based on other species. Another attraction of the candidate gene approach is that unlike quantitative genetic approaches, it does not require complicated breeding designs or data on the relationships among individuals – comparisons can be made across individuals of unknown relatedness because the genes are being measured directly. Therefore this approach is suitable for organisms that do not readily breed in the lab.

Summaries of using the candidate gene approach for studying behaviour are well-described elsewhere. Notable examples include work on the *for* gene and foraging behaviour, and the 'fruitless' gene and mating behaviour in Drosophila and the promoter region of the vasopressin gene and parental behaviour. The candidate gene approach for studying the genetics of fish behaviour is a relatively unexplored but promising future research direction. While the candidate gene approach is very attractive and has great appeal especially for non-model organisms, there are some drawbacks. First, candidate gene studies are biased; what if the ultimate source of genetic differences between groups lies further up or downstream of the particular candidate gene? One approach to this issue is to choose a pathway that is probably associated with the behaviour and look at gene expression at several points in the pathway. Second, candidate gene studies are purely correlative. Concluding that the gene is really associated with the behaviour requires further experimentation. Perhaps most seriously, candidate gene approaches work best when the trait is influenced by just few genes of major effect, which is probably the exception rather than the rule for behavioural traits.

GENOMICS

'Genomics' refers to the study of the structure, content and evolution of genomes, including the analysis of the expression and function of genes and proteins. What distinguishes genomics from other branches of genetics is that it looks at the whole genome simultaneously, rather than focusing on one gene at a time. Luckily, doing genomics does not require a full genome sequence, and there are already several good examples of applying whole-genome approaches to non-model organisms and ecologically-relevant traits, including behaviour. Genomics has been hailed as an opportunity to integrate mechanistic and evolutionary approaches to studying behaviour because information about genes provides neuroscientists and behavioural ecologists with a 'common language'.

Interspecific Hybridisation

Interspecific hybridisation has been used to increase growth rate, manipulate sex ratios, produce sterile animals, improve flesh quality, increase disease resistance, improve tolerance of environmental extremes, and improve a variety of other traits that make aquatic animal production more profitable. Although interspecific hybridisation rarely results in an F1 suitable for aquaculture application, there are a few significant exceptions. Channel catfish females x blue catfish, I. furcatus, males ("channel-blue") is the only cross (between 28 catfish hybrids examined) that exhibits superiority for growth rate, growth uniformity, disease resistance, tolerance of low oxygen levels, dressing percentage and harvestability. However, mating problems between the two species have prevented commercial utilisation. The "sunshine" bass is a cross between white bass, Morone chrysops, and striped bass, M. saxatilis, and grows faster, with better overall culture characteristics than either parent species. In addition, crosses of the silver carp (Hypophthalmichthys molitrix) and bighead carp, black crappie and P. annularis and African catfish hybrids all show faster growth than parent species. Numerous crosses of common carp with rohu, mrigal and catla; tambaqui and Piaractus brachypoma and P. mesopotamicus green sunfish crossed with bluegill, and gilthead seabream with red seabream, have also resulted in improved overall performance for aquaculture systems. In the family Sparidae, hybrids of P. major and common dentex,Dentex dentex, also grow faster than parent stocks.

Hybridisation in commercial Thai oysters was carried out to explore the possibility of producing hybrid oysters with superior traits. Hybridisation of C. belcheri × C. lugubris was only successful to the spat stage, and growth rates of the hybrids and reciprocal crosses were significantly lower than their parents. Intergeneric hybridisation was only successful with female C. lugubris and male S. cucullata. Growth rates of this hybrid were significantly higher than those of S. cucullata, but did not differ significantly from those of C. lugubris. Shell morphology of the hybrid was intermediate between the two parental types. Effects of intergeneric hybridisation on the genetic diversity of natural oyster populations.

Hybridisation between some species, such as Nile tilapia and blue tilapia, Oreochromis aureus, result in predominantly male offspring. Other tilapia crosses, which produce mainly male offspring, include Nile tilapia × O. urolepis honorum or O. macrochir, and O. mossambicus × O. urolepis honorum. Conversely, the cross between striped bass and yellow bass produced 100 per cent females. This can be desirable for culture purposes where i) there are: growth differences between the sexes; ii) sex-specific products such as caviar are wanted; or iii) reproduction needs to be controlled (*e.g.*, overpopulation and stunting in tilapia production ponds).

Hybridisation between species can also result in offspring that are sterile or have diminished reproductive capacity. As with monosex production, the production of sterile hybrids can reduce unwanted reproduction or improve growth rate by energy diversion from gametogenesis. Karyotype analysis can be used as a general predictor of potential hybrid fertility. For example, hybrid Indian major carps are generally fertile because they share similar chromosome numbers (2N = 50). When crossed with common carp (2N = 102), the hybrids are triploid and sterile. A natural triploid results from the cross between grass carp, Ctenopharyngodon idellus, and bighead carp. Grass carp are commonly produced for aquatic weed control, but there is concern about spread to natural water bodies and the potential impact on desirable vegetation. The triploid hybrids have reduced fertility, but some progeny maintain diploidy and could be fertile. Other exceptions to the chromosome number-fertility rule are crosses of some sturgeon species with different chromosome numbers that produce fertile F1 offspring. Investigated hybridisation of coho salmon, which is considered resistant to several salmonid viruses. Disease resistance in the hybrids was improved, but overall viability was poor.

Viability increased when hybridisation was followed with triploidisation. The same authors also reported that triploid hybrids from rainbow trout and char were resistant to several salmonid viruses, but grew more slowly than their diploid counterparts. Similar results were found with rainbow trout and coho crosses. Hybrid triploids of Atlantic salmon x brown trout showed survival and growth rates comparable to those of Atlantic salmon monospecific diploids. Triploid Pacific salmon hybrids have also shown earlier seawater acclimation.

Increased environmental tolerance may also be imparted to hybrids where one parent species has a wide or specific physiological tolerance or due to increased heterozygosity. Hybrid red tilapia, O. mossambicus, (high salinity tolerance) and Nile tilapia, O. niloticus, (low salinity tolerance) show enhanced salinity tolerance. Florida red-strain hybrids can reproduce in salinities of 19 ppt and O. niloticus × O. aureus hybrids also show enhanced salinity tolerance. Reciprocal hybrids of O. niloticus (N) × O. mossambicus (M) have different salinity tolerances.

The hybrid with an O. niloticus mother (N×M) had a higher survival rate after salinity challenges at 20 ppt than pure O. niloticus, but lower survival rates than those of the reciprocal hybrid (M×N). At 30 ppt salinity, a direct transfer killed all tilapia with O. niloticus maternal genes. Growth rates of N×M hybrids were comparable to those of Nile tilapia, while those of the N×M hybrids and O. mossambicus were comparable, but lower, than the first two groups. There were no statistical differences in carcass percentages of the four groups. Back-crosses were also evaluated. MN×N showed the highest salinity tolerance (comparable to that of O. mossambicus), but no significant differences in salinity tolerances were found in the remaining backcross (N×NM, NM×N, N×MN) or pure O. niloticus. Carcass percentage of the back-cross hybrids, however, tended to be higher than those of the parent species.

Hybrids between marine species, and between marine and freshwater spawning species, have produced surprising results. Hybrids between Sparus aurata and Pagrus major (both belonging to the Sparidae) developed vestigial gonads at two to three years and were sterile. Similar vestigial gonads were observed in offspring of the reciprocal crosses. No consistent growth or survival superiority, compared with the parent species, was detected until sexual maturity in the reciprocal crosses. Hybridisation between European sea bass, females and striped bass resulted in viable larvae; however 28 per cent were triploids, and only the triploids appeared to survive to six months of age. At eight months, the survivors showed poor growth compared to diploid D. labrax.

Such hybrids may be of commercial value where reproduction needs to be restricted for ecological reasons.

Sometimes an interspecific hybrid does not exhibit heterosis for specific traits, but may still be important for aquaculture if it expresses other useful traits from the parent species. The main catfish cultured in Thailand is the hybrid between African (Clarias gariepinus) and Thai catfish. This combines the fast growth of the African catfish and the desirable flesh characteristics of the Thai catfish. Although it does not grow as fast as pure African catfish, overall production rates are improved and the flesh is still acceptable to Thai consumers. Likewise the rohu x catla hybrid grows almost as fast as pure catla, but has the small head of the rohu considered desirable in Indian aquaculture. Catla catla × Labeo fimbriatus (fringe-lipped peninsula carp) hybrids have the small heads of L. fimbriatus, plus the deep body and growth rate of catla. The "sunshine" bass has a suite of advantageous traits from the parent species (white and striped bass) including good osmoregulation, high thermal tolerance, resistance to stress and certain diseases, high survival under intense culture, and ability to use soy bean protein in feed. Interspecific backcrossing has also been used to successfully introgress (merge) genes from one closely related species into another. Cold tolerance and colour genes have been transferred among tilapia in this manner.

Genetic Selection

Very little was done in this area prior to 1970, however, the field has grown significantly in the past three decades and is now extremely active. In general, the response to selection for growth rate in aquatic species is very good compared to that with terrestrial farm animals. Fish and shellfish often have higher genetic variance compared to farmed land animals, notes the genetic variation for growth rate is seven to ten per cent in farmed land animals and 20-35 per cent in fish and shellfish. Fecundity is also higher in aquatic organisms than land organisms. This allows for higher selection intensity for aquaculture production improvement, and over 200 heritability estimates have been obtained for several traits of cultured fish and shellfish.

There are few national breeding programmes in fish and shellfish which aim at improved aquaculture production. In 1975, a National Breeding programme for Atlantic salmon and rainbow trout was started in Norway, and today this supplies about 75 per cent of the Norwegian industry with improved eyed eggs. In Canada, a similar breeding programme is operated by the Atlantic

Salmon Federation. In 1993, The Philippines National Tilapia Breeding programme (PNTBP) was started with broodstock from the GIFT programme (Genetically Improved Farmed Tilapia) and is now operated by the GIFT Foundation. In Israel and Hungary, crossbreeding programmes with common carp (Cyprinus carpio) exist. In addition, some private companies in several countries, including the United States and Chile, now have their own breeding programmes.

Selection for body weight and disease resistance in salmonids has been particularly successful. By 1925, three generations of selection of brook trout (Salvelinus fontinalis) survivors of endemic furunculosis (caused by Aeromonas salmonicida) improved survival from 2 per cent to 69 per cent. Ehlinger further increased resistance to furunculosis in brown trout and brook trout with selective breeding programmes. An infectious pancreatic necrosis virus (IPNV)-resistant strain of rainbow trout showed 4.3 per cent mortality compared with 96.1 per cent in a highly sensitive strain. With respect to body weight, a 30 per cent increase in rainbow trout was achieved within six generations of selection. In Atlantic salmon, an increase of seven per cent was achieved within a single generation and an increased growth rate of 50 per cent was achieved within ten generations in coho salmon. Body weight has also been improved in channel catfish, by 12-20 per cent over one to two generations of genetic selection. The best line grew twice as fast as non-selected lines.

After three generations, the growth rate of channel catfish in ponds was improved by 20-30 per cent. Four generations of selection in a Kansas strain of channel catfish resulted in 55 per cent improvement in growth rate. Four generations of selection also increased body weight by 50.5 gm and total length by 0.88 cm in walking catfish, C. macrocephalus. Genetic selection in gilthead seabream (Sparus aurata) has also been successful despite early difficulties with producing single-pair offspring groups. This led to the conclusion that family mating designs were inappropriate for group spawning of S. aurata. Mass selection proved more effective and resulted in significant heritability estimates for growth. Different strains of common carp appear to possess varying amounts of additive genetic variation. Heritabilities for body weight of 0.15-0.49 in a Czechoslovakian strain of common carp. Vietnamese common carp also show significant heritability (0.3) for growth rate.Successful selection programme, which started in 1965, against dropsy (spring viremia of carp) in common carp.

Responses can differ depending on the direction of selection. Body weight of common carp in Israel was not improved over five generations, but could be decreased in the same strain selected for small body sise. Virtually identical results for Nile tilapia has also been reported.

Several authors reported that, in tilapia, mass selection improved body weight in Oreochromis mossambicus, red tilapia and O. aureus. However, selection for increased body weight in other red tilapia has been less successful.

This is similar to the situation in Nile tilapia. This may reflect a narrow genetic base in the founder stock or sole use of mass selection. Selection for increased growth in GIFT Nile tilapia gave much different results, with 77 per cent to 123 per cent growth improvement. The genetic gain was superior to results from crossbreeding experiments.

The 11 per cent genetic gain per generation in GIFT tilapia is better than that obtained in most other species of fish; the channel catfish increased body weight by 14 per cent per generation over four generations. However, a more typical genetic gain is five to seven per cent per generation, as demonstrated by salmonids following approximately ten generations of selection. The only other exceptions that come close to the results with GIFT tilapia and the channel catfish study of Padi are the 13-14 per cent increase per generation observed by Gjerde, and five per cent per generation for common carp after six generations of selection.

Production trials and socio-economic surveys in five Asian countries show that cost of production per unit fish is 20-30 per cent lower for GIFT strain tilapia than other Nile tilapia strains in current use.

Body weight of common carp appears unresponsive to selection; however, body conformation can be dramatically changed. Selection for carcass quality and quantity has also been initiated for salmonids and catfish. In Thailand, selection results are not yet available for many species and traits, but numerous heritability estimates have been obtained, *e.g.*, for growth and disease resistance in pangasiid freshwater catfish, rohu (Labeo rohita), Thai walking catfish (Clarias macrocephalus), Java barb (Barbodes gonionotus), bighead carp (Aristichthys nobilis) and Asian rock oyster (Saccostrea cucullata).

Selective breeding has also improved growth of the shrimp Penaeus japonicus in laboratory and pilot-scale farm trials using offspring from CSIRO broodstock. In 1998/1999, comparative trials demonstrated significant improvements in the growth, survival and total yields in two selected lines (10-15 per cent increase in mean yields). In a related species, P. vannamei, estimated a response within one generation of selection of 4.4 per cent for growth rate and 12.4 per cent for survival of the viral disease Taura syndrome. A genetic improvement programme was recently started for Pacific oysters, Crassostrea gigas, in Australia. This followed demonstrations that little genetic diversity had been lost since the Pacific oyster industry was founded in Australia with imports from Japan some 50 years ago. The improvement programme aims to combine family and mass selection with molecular genetics. Two generations of mass selection and two generations of family selection have been completed, with a growth rate improvement of about eight per cent in the first generation from a mass selection. Based on work with a congeneric oyster species, C. virginica, Mass selection of adult oysters gave an apparent strong response to selection for growth rate.

CELL CULTURE IN AQUACULTURE

Development and characterization of cell lines from selected finfish and shell fish was undertaken at CIFE, Mumbai and a cell culture system was developed from the fish Lates calcarifer (Sea bass). At CUSAT, the cell lines developed from eyestalk sub-cultured and made available to develop such cell lines as and when required. The cells shown differentiation in to a proliferating bunch of cells and subsequently got detached as a lump of cell aggregates. The development of cell lines from seabass, Lates calcarifer was attempted at CAH Abdul Hakeem College, Melvisharam and CUSAT. The kidney cells shown fibroblastic and formed a monolayer. The cells were were deposited at NCCS.

Growth Promoting Factor(s)/Signal Transduction

A collaborative project was undertaken on identification, isolation and characterization of growth promoting and differentiation factor(s) from perivitelline fluid of developing embryos of the Indian horseshoe crab at Agharkar Research Institute, Pune, National Centre for Cell Science (NCCS), Pune and National Institute of Oceanography (NIO), Goa. Active fraction identified shown significant enlargement and enhanced compartmentalization of the heart in cardiac development and promoting activity. In another project on large-scale cultivation of amoebocytes in vitro from Indian horseshoe crab implemented by NCCS and NIO, regeneration of gill lamellae of the Indian horseshoe crab was successful in amoebocyte production. A molecular signal intervention mechanism was developed as an alternative to eyestalk ablation at the Departments of Zoology and Biochemistry, University of Madras. A specific inhibitor that intervenes in this pathway has been shown to reduce the molt cycle of the prawn and reduces the turnover time and rearing cost and maximizes the economic returns. By decreasing the molt cycle duration, it can accelerate the growth rate of the prawn resulting in early maturation.

Studies on feed supplementation of highly unsaturated fatty acids (HUFAs) and immunostimulants for disease resistance in freshwater prawn was carried out at the Department of Aquaculture, College of Fisheries, Mangalore. HUFAs namely Eicosapentaenoic acid (C20: 5n-3) and Docosahexaenoic acid (C22: 6n-3) are important for larval growth. The supplementation of PUFA incorporated feed could enhance growth, survival and disease resistance in freshwater prawn. A project on prototyping of raceway based third generation shrimp production technology was implemented at the Fisheries Biotechnology Centre, Fisheries College and Research Institute, Tuticorin. The protocol has been developed to increase survival of post-larvae. This in-door shrimp farming facility will be eco-friendly, economically viable and technically useful for large-scale adoption in aquafarming systems in India. The work on development of Artemia franciscana culture in the solar salt ponds and processing of Artemia cyst and biomass was undertaken at Centre for Marine Science and Technology, Manonmaniam

Sundaranar University, Rajakamangalam, Tamil Nadu. Artemia production was initiated in condenser ponds and inoculated in culture ponds and the efforts are being made to utilize the abandoned area of the saltworks for Artemia cultivation. Studies were undertaken at NIO, Goa to isolate, purify and characterize new natural fluorescent dyes and fluorescent molecules from tissues of sea cucumber Holothuria scabra and to test them as bioactivities compounds for therapeutic and industrial applications. At Kerala Agriculture University, Kochi, attempts were made to isolate bioactive substances such as antibacterial, antiviral and anticancer agents from the accessory nidamental gland and ink of cuttlefish (Sepia pharaonis ehrenberg). An antitumour peptidoglycan was isolated from cuttlefish ink and the crude preparation showed antitumour activity against mice following intraperitonial administration. The peptidoglycan was made up of protein and uronic acid rich with polysaccharide fraction and was found to be made up of five amino acids. The screening of marine Acinetobacter genospecies for production of biosurfactant/s, purification and antimicrobial activity was undertaken at Pune University. Untapped natural sources were screened for novel biosurfactant/s and bioemulsifier-producing microorganisms using Acinetobacter strains, which showed emulsification of toluene, diesel, hexadecane and crude oil, haemolysis of human RBC's and anionic biosurfactant/s production.

At the department of biotechnology - CUSAT, the work on molecular cloning was undertaken to develop a technology for production of alkaline protease enzyme using marine fungus, Engyodontium alba. The optimal conditions were studied for enzyme production using solid-state fermentation of wheat bran as the solid substrate. The enzyme was found relatively stable for one year. At NIO, Goa, fungi from deep-sea sediments were explored as new sources for protease. Fungal growth and protease production at low temperature and elevated hydrostatic pressure was studied to understand their adaptations to extremophilic conditions.

SEAWEED CULTIVATION

At CSMCRI, Bhavnagar, the project was undertaken on scale-up cultivation and processing of phycocolloid seaweeds (Gelidiella acerosa and Eucheuma). Multilocational trials undertaken to validate different culture methods for large-scale cultivation of Eucheuma proved that the raft cultivation method consistently yields better harvest. This method has been successfully demonstrated in the Gulf of Mannar, Tamil Nadu coast as well as at Okha, Gujarat coast. High quality agar was extracted from Gelidiella acerosa and Gracilaria sp. The Gracilaria agar had very high gel strength (1200 g cm-2) with low gelling temperature. This agar will be useful in bacteriological and molecular biology applications.

The nitrifying bioreactor technology developed at CUSAT paves the way for transforming the open prawn/shrimp larval rearing systems to closed recirculation systems. The reactor, on properly activating and deploying in the hatchery system, can take care of the entire water quality problems. Apart from nitrification this duel capability lowers down the biological oxygen demand, total bacterial population along with stabilization of Vibrio population. The training programme for fisheries scientists was undertaken at three institutes namely Central Institute of Freshwater Aquaculture (CIFA), Bhubaneshwar, Central Institute Fisheries Technology (CIFT), Cochin and Central Institute of Fisheries Education (CIFE), Mumbai to train in service fisheries scientists on aspects of modern biology, health management and conservation of resources etc. Trainees are being given both theoretical hands-on training and independent project work on molecular biology aspects relevant to fisheries topics. After successful completion of the training programme, the trainees have started working in the area of molecular biology in marine sciences at their institutes.

New Programme Support

The programme on aquaculture and marine biotechnology has been identified for support after a detailed discussion under the Task Force on Aquaculture and Marine Biotechnology. Two proposals from the UAS, College of Fisheries, Mangalore and the Cochin University of Science and Technology (CUSAT), Cochin have been supported for attention on brood stock and larval health management of black tiger shrimp and freshwater prawn and screening for potential anticancer agents from marine Cnidarians, bacteria and fungi. Genotypic characterization and sequencing of genes of interest of marine bacteria, fish and shellfish immunology and screening for useful natural products from microorganisms, capacity building through hands-on training programmes for scholars and technicians on molecular techniques for detection of pathogens and health management in shrimp hatchery system, microbial taxonomy, crustacean immunology and screening and characterization of marine microbes for useful products etc. are also the part of the programme.

Shrimp Genomics

Shrimp (Penaeus monodon) is a very important species for the country from the commercial viewpoint. Unfortunately, the species is not domesticated and the harvests are mainly from the wild. The reproductive biology is complex and the disease burden very high, especially viral diseases. There are populations, which are naturally resistant to viral attack and could be tapped to our advantage. The network programme on shrimp genomics has been developed covering the aspects *viz.* functional genomics in disease resistance of Penaeus monodon, gene regulation of maturation and breeding in P. monodon, microsatellites and physical mapping of P. monodon and genomics of Penaeus indicus for comparative genomics.

Bibliography

A.C. Long: *A-Z of Industrial Marine Fisheries in Asia*, Cyber Tech Publications, New Delhi, 2011.

A.D. Dholakia: *Fisheries and Aquatic Resources of India*, Daya Publication, Delhi, 2004.

Ajit Kumar Roy and Niranjan Sarangi: *Applied Bioinformatics Statistics and Economics in Fisheries Research*, New India Publication Agency, Delhi, 2008.

Ajit Kumar Roy: *Evaluation and Impact Assessment of Technologies and Developmental Activities in Agriculture, Fisheries, and Allied Fields*, New India Publishing Agency, Delhi, 2011.

B K Singh: *Applied Fisheries and Aquaculture*, Swastik Publication, Delhi, 2008.

B. N. Pandey and G.K. Kulkarni: *Fisheries and Fish Toxicology*, APH Publication, Delhi, 2011.

B.N. Pandey, Sunil P. Trivedi, Kamal Jaiswal and Navneet Kaur Sethi: *Fish and Fisheries*, Sarup Book Publishers, Delhi, 2009.

C.B.L. Srivastava: *A Text Book of Fishery Science and Indian Fisheries*, Kitab Mahal, Delhi, 2013.

Gary T. Sakagawa: *Assessment Methodologies and Management: Proceedings of the World Fisheries Congress, Theme 5*, Oxford & IBH, Delhi, 1995.

H.R. Singh and W.S. Lakra: *Coldwater Aquaculture and Fisheries*, Narendra Publication, Delhi, 2000.

Hrishikes Bhattacharya: *Commercial Exploitation of Fisheries : Production, Marketing, and Finance Strategies*, Oxford University Press, Delhi, 2002.

J.K. Roshan: *A-Z Fish and Fisheries*, Centrum Press, Delhi, 2009.

J.S. Datta Munshi, J. Ojha and T.K. Ghosh: *Advances in Fish Research : Vol: 3: Fisheries and Fish Biology Research*, Narendra Publication, Delhi, 2003.

K. C. Pandey: *Concepts of Indian Fisheries*, Shree Publishers , Delhi, 2012.

K.D. Bhardwaj: *Business Management in Fisheries and Aquaculture*, Cyber Tech Publication, New Delhi, 2012.

K.P. Biswas: *Advancement of Fish Fisheries and Technology*, Narendra Publishing House, Delhi, 2012.

K.P. Biswas: *Ecological and Fisheries Development in Wetlands : A Study of Chilka Lagoon*, Daya Publication, Delhi, 1995.

K.P. Biswas: *Economics in Commercial Fisheries*, Daya Publication, Delhi, 2006.

Maniranjan Sinha: *Fish and Fisheries of North-Eastern States of India*, Narendra Publishing House, Delhi, 2011.

Nasreen Akhter: *Economics in Fisheries Research*, Oxford Book, Delhi, 2010.

Neha Charan: *Classification of Fisheries*, Random Publications, Delhi, 2013.

P N Panday; B K Gorai and B C Jha: *Economics of Fisheries : Ecology and Fisheries of Oxbow Lakes*, APH Publication, Delhi, 2005.

Rajendra Kumar: *A Textbook of Fish and Fisheries*, Arise Publication, Delhi, 2008.

Rajendra Kumar: *Biotechnology and Genetics in Fisheries and Acquaculture*, Arise Publication, Delhi, 2010.

Santosh Kumar and Manju Tembhre: *Fish and Fisheries : Anatomy, Physiology, Applied Fisheries, Genetics, Biotechnology, and Fish Legislature*, New Central Book Agency (P) Ltd, Delhi, 2010.

V P Saini; L L Sharma and N C Ujjania: *Dictionary of Aquatic Resources and Fisheries*, Agrotech Books, Delhi, 2007.

V. Suryanarayan: *Conflict Over Fisheries in the Palk Bay Region*, Lancer Publication, Delhi, 2005.

V.B. Sakhare and B. Vasanthkumar: *Emerging Trends in Fisheries and Aquaculture*, Daya Publication, Delhi, 2013.

Varun Mehta: *Fisheries and Aquaculture Biotechnology*, Campus Books International, Delhi, 2006.

Index

I

L

M

N

O

P

Q

R

S

T

U

V

W